# Design Against Blast

## Load Definition & Structural Response

## WITPRESS

WIT Press publishes leading books in Science and Technology.
Visit our website for the current list of titles.
www.witpress.com

## WITeLibrary

Home of the Transactions of the Wessex Institute, the WIT electronic-library
provides the international scientific community with immediate and permanent
access to individual papers presented at WIT conferences.
Visit the WIT eLibrary athttp://library.witpress.com

# Design Against Blast

## Load Definition & Structural Response

**Edited by**

**S. Syngellakis**
*Wessex Institute of Technology, UK*

WIT*PRESS* Southampton, Boston

Editor:

**S. Syngellakis**
*Wessex Institute of Technology, UK*

1006917112

Published by

**WIT Press**
Ashurst Lodge, Ashurst, Southampton, SO40 7AA, UK
Tel: 44 (0) 238 029 3223; Fax: 44 (0) 238 029 2853
E-Mail: witpress@witpress.com
http://www.witpress.com

For USA, Canada and Mexico

**WIT Press**
25 Bridge Street, Billerica, MA 01821, USA
Tel: 978 667 5841; Fax: 978 667 7582
E-Mail: infousa@witpress.com
http://www.witpress.com

British Library Cataloguing-in-Publication Data
A Catalogue record for this book is available
from the British Library

ISBN: 978-1-84564-750-6
eISBN: 978-1-84564-751-3

Library of Congress Catalog Card Number: 2012949842

# Preface

Terrorist attacks and other destructive incidents caused by explosives have, in recent years, prompted considerable research and development into the protection of structures against blast loads. For this objective to be achieved, experiments have been performed and theoretical studies carried out to improve our assessments of the intensity as well as the space-time distribution of the resulting blast pressure on the one hand and the consequences of an explosion to the exposed environment on the other.

This book aims to enhance awareness on and understanding of these topical issues through a collection of relevant, Transactions of the Wessex Institute of Technology articles written by experts in the field. The book starts with an overview of key physics-based algorithms for blast and fragment environment characterisation, structural response analyses and structural assessments with reference to a terrorist attack in an urban environment and the management of its inherent uncertainties.

A subsequent group of articles is concerned with the accurate definition of blast pressure, which is essential prerequisite to the reliable assessment of the consequences of an explosion. A variety of computer codes, associated with different types of explosive charges and ranges, exist for this purpose. These models range from simple empirically based to multi-physics, high strain rate finite element implementations of the conservation equations of continuum mechanics. Articles in the book address the experimental validation of available software as well as the comparison of their blast profile predictions.

Blast pressure profiles and histories are affected by the geometric configuration of the exposed structure and this particular issue is dealt with in the case of a round column. In the context of continuum modelling, the equation of state for the detonation products describes the work output from the explosive that causes the subsequent air blast. The reliable parameter calibration for this equation is an important topic also covered in the book.

Other papers are concerned with alternative methods for the determination of blast pressure. These are based on experimental measurements or neural networks. The latter model was trained and validated using as input blast threat scenarios and exposed geometry parameters and as output CFD predictions of peak pressures and impulses.

Numerical, theoretical and experimental studies on the effect of blast on components, structures and its occupants combine physics of detonation chemistry, shock physics, solid mechanics, structural dynamics, nonlinear material behaviour, human physiology and injury mechanics. A number of studies address general issues such as the generation of laboratory simulation of blast that reproduces failure modes of reinforced concrete elements observed in field tests, the link between non-ideal explosive and target configuration and the coupling of the response to the blast loading. The latter can be strong, particularly for close-in blast loading configurations; its effect is investigated in the case of flexible systems such as membranes, blast curtains, cable facades as well as compressible, soil-filled, concertainer walls. A final group of articles reports investigations on predicting the response of specific structural entities and their contents. Simplified methods for buildings, either taken from existing literature or based on experimental and numerical results are compared; another simplified method based on a beam-column damage element is tested on portal frames, which include laminated composite beams. Bridges have been given considerable attention as potential transportation targets; one paper describes a methodology for generating detailed design guidelines and another, an approximate method for the response of long span highway girders. Two articles on the protection of vehicles against explosive devices, such as mines and improvised explosive devices, put particular emphasis on occupant safety. The book concludes with studies on the effectiveness of steel-reinforced polymer in improving the performance of reinforced concrete columns and the failure mechanisms of seemless steel pipes used in nuclear industry.

Stavros Syngellakis (ed.)
The New Forest, 2013

# Acronyms

| | |
|---|---|
| ALE | Arbitrary Lagrangian- Eulerian |
| AMR | Adaptive Mesh Refinement |
| ANFO | Ammonium Nitrate/Fuel Oil |
| ANN | Artificial Neural Network |
| AWE | Atomic Weapons Establishment |
| BG | Blast Generator |
| CAD | Computer Aided Design |
| CCD | Charge-coupled Device |
| CFD | Computational Fluid Dynamics |
| CFRP | Carbon Fibre Reinforced Polymer |
| CJ | Chapman Jouguet |
| CMU | Concrete Masonry Unit |
| DLF | Dynamic Load Factor |
| DOT | Department of Transportation |
| DRDC | Defence R&D Canada |
| DTRA | Defense Threat Reduction Agency |
| EFP | Explosively Formed Penetrator |
| EMRTC | Energetic Materials Research and Testing Center |
| EOS | Equations of State |
| FSI | Fluid Structure Interaction |
| GIS | Geographical Information System |
| GMDH | Group Method of Data Handling |
| GRNN | General Regression Neural Network |
| GSA | General Services Administration |
| GUI | Graphical User Interface |
| HE | High Explosive |
| HIC | Head Injury Criterion |
| ICP | Integrated Circuit Piezoelectric sensor |
| IED | Improvised Explosive Device |
| IMEA | Integrated Modular Effectiveness Analysis |
| IRSN | Institut de Radioprotection et de Sûreté Nucléaire (Radio protection and Nuclear Safety Institute) |

| | |
|---|---|
| JWL | Jones-Wilkins-Lee |
| LSTC | Livermore Software Technology Corporation |
| LVDT | Linear Variable Differential Transformer |
| MEVA | Modular Effectiveness/Vulnerability Assessment |
| MLP | Multi-Layer Perceptron |
| NCAP | New Car Assessment Programme |
| NCHRP | National Cooperative Highway Research Program |
| PETN | Pentaerythritol tetranitrate |
| P-I | Pressure-impulse |
| PILR | Propagation, Interaction, Load and Response |
| RC | Reinforced Concrete |
| RHA | Rolled Homogenous Armour |
| SBD | Simulation-Based Design |
| SPH | Smooth Particle Hydrodynamics |
| SRP | Steel Reinforced Polymer |
| TNT | Trinitrotoluene |
| TROSS | Test Rig for Occupant Safety Systems |
| USACE | United States Army Corps of Engineers |
| ZND | Zel'dovich-von Neumann-Doring |

# Contents

# Integrated anti-terrorism physics-based modelling: threats, loads and structural response

F. A. Maestas, J. L. Smith & L. A. Young
*Applied Research Associates Inc., USA*

## Abstract

Modelling of a terrorist attack in an urban environment requires a balanced understanding of the complex physical processes that occur and management of the inherent uncertainties associated with the modelling. This paper examines the key physics-based techniques required to accurately model a terrorist attack in an urban environment. Specifically, this paper will address the blast and fragment environment that results from an Improvised Explosive Device (IED) detonation, the loads on an urban structure and the response of that structure, to include progressive collapse. Tools such as the United States Air Force Research Laboratory's (AFRL's) Modular Effectiveness/Vulnerability Assessment (MEVA) code and the Defense Threat Reduction Agency's (DTRA's) Integrated Modular Effectiveness Analysis (IMEA) tool embody algorithms for blast and fragment environment characterization, structural response analyses, and structural assessments. The key physics-based algorithms in these tools and others will be highlighted. Additionally, this paper will provide an approach to handling the uncertainties of modelling the urban structure given limited knowledge of the building's key structural attributes. The combination of treating uncertainties in a physics-based approach provides an integrated modelling method for evaluating and planning for a terrorist attack.
*Keywords:    weapon effectiveness, survivability analysis, modelling and simulation, physical security analysis, personnel security.*

## 1   Introduction

Modelling of a terrorist attack in an urban environment requires a balanced understanding of the complex physical processes that occur and management of the inherent uncertainties associated with the modelling. The process used in most modelling tools can be summarized with a simple acronym: PILR; Propagation, Interaction, Load and Response. This paper will discuss the PILR model as it applies to Improvised Explosive Devices (IED) in urban environments. The focus of the paper will be associated with blast and fragment aspects. Figure 1 provides an illustration of this concept. Propagation is the environment that results from the detonation of an IED. Interaction describes how the environment interacts with the urban structure. Load refers to the load on the structure which results from the environment. Response is how the structure and contents, to include equipment and people, respond to that environment.

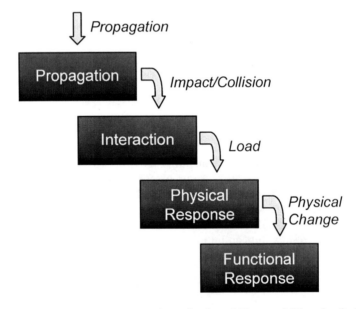

Figure 1:    PILR representation of vulnerability modelling for facility.

## 2  Propagation

An IED detonation environment can be divided into two primary aspects, blast and fragment.

### 2.1 Blast

The level of fidelity in blast models varies somewhat from code to code. Most weapon effectiveness or survivability models provide analytical approximations for the shock(s) that result from the detonations. These blast pressure time histories for both the static (side-on) pressure and dynamic pressure environments are evaluated. Figure 2 provides a simplified method for obtaining peak free field pressure as a function of scaled range. The scaled range is defined as the distance from the detonation to the point divided by the cube root of the effective explosive weight. Equation (1) provides a method for calculating the pressure time history. The peak pressures, time histories and the integration of the time history (impulse, fig. 3) are used as loads on the structure, equipment and inhabitants. These blast models are generally only appropriate for conventional high explosives and are used to generate the ideal, free-field environment [3, 4].

$$P(t) = P_{op}\left(1 - \frac{t - t_1}{t_o}\right)e^{-(t-t_1)Do} \tag{1}$$

where $P_{op}$ is the peak pressure, $t$ the time, $t_1$ the shock arrival time, $t_o$ the time of positive phase and $D_o$ is a decay constant.

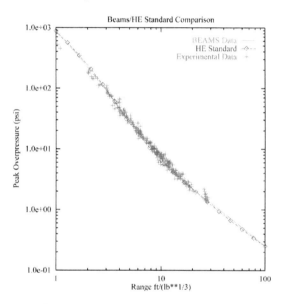

Figure 2:    Free field peak pressure.

Figure 4 provides the peak dynamic pressure, density, and particle velocity as a function of the peak free field pressure. The dynamic pressure, also known as "gust," is the pressure caused by the motion of the gas, which is equal to $(1/2)\rho U^2$, where $\rho$ is the gas density and $U$ is the gas velocity. The dynamic pressure is sometimes referred to as "differential pressure."

Since most IEDs are not made from TNT, the equivalent explosive weight, $W_e$, is calculated by scaling the energy to the explosive of interest.

## 2.2 Fragments

IEDs used to be employed by Iraqi insurgents at a rate of approximately 40 per day. Many IEDs are constructed using unexploded inventoried ordnances. Thus, fragment fly-out can be modelled using a stochastically generated set of weapon fragments, based upon either Arena test data files or Mott's distribution. Simplified algorithms as provided in eqns. (2)–(5) can be used. Figure 5 provides an indication of fragment density as a function of fragment size.

$$V = V_G \left[ 1 - \left( \frac{R_f}{R_o} \right)^{2(1-\gamma)} \right]^{\frac{1}{2}}$$

(2)

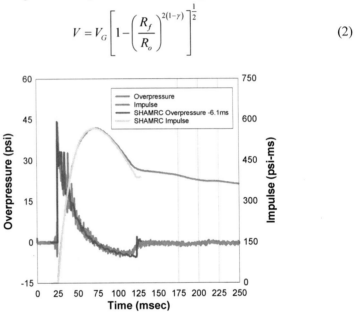

Figure 3:    Example of calculated pressure and impulse time histories.

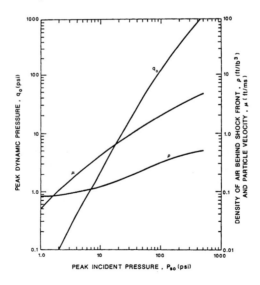

Figure 4:    Peak dynamic pressure, density and particle velocity.

$$V_G = \sqrt{2E} \left[ \frac{M}{C} + \frac{\left(1/2 - \alpha/3 - \alpha^2/6\right)}{\left(1 - \alpha^2\right)} \right]^{\frac{1}{2}} cylinder \qquad (3)$$

where $M$ is the mass of the casing, $C$ the mass of explosive, $E$ the Gurney energy and $\alpha = r/R$.

$$n = \lambda e^{-\lambda y}$$
$$m = y \lambda e^{-\lambda y} \qquad (4)$$

$$N = \int_O^Y \lambda e^{-\lambda y} dy$$
$$M = \int_O^Y y \lambda e^{-\lambda y} dy \qquad (5)$$

where $n$ is the fragment density (normalized), $m$ the mass density (normalized), $N$ the total fragments (normalized), $M$ the total mass (normalized),

$$\lambda = \frac{1}{mean\ fragment\ mass}$$

and $y$ the fragment size variable.

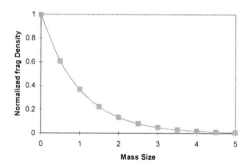

Figure 5:    Example of calculated fragment size distribution interaction.

For *ad hoc* IED devices, fragments in a terrorist environment may be pieces of a disassembling vehicle (for a vehicle borne IED), or shards of glass or nuts and nails (for a suicide bomb). Those fragment models do not exist. The US Army Research Laboratory and others are working to address this shortfall; however, in the interim, the use of Arena data from an inventory weapon would provide an upper bound of the damage caused by fragments.

## 3   Interaction

Key to modelling the interaction of environment with the structure is accurately modelling the urban structure. This requires more than a Computer Aided Design (CAD) representation. Physical properties of the elements, such as compressive strength and their connectivity, are also required. MEVA for example, uses a 3D Complex Solid Modelling tool named the Smart Target Model Generator to add strength parameters to the facility model.

### 3.1  Blast

The free field blast environment must be adjusted based upon where the detonation occurs in relation to hard surfaces, fig. 6. For example, if the detonation occurs within a distance of $1.5W^{1/3}$, then the peak pressure is multiplied by a factor of 1.8. This results in an effective explosive weight, $W_e$, of $1.8W$. If near a corner (two surfaces), then the factor is 1.8×2 [5].

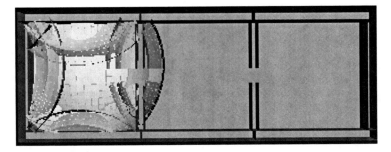

Figure 6:   Example of internal blast reflection and propagation inside building.

### 3.2  Fragment

The fragment environment is typically modified by ricochet or velocity reduction because of the energy loss that occurs due to penetration. Ricochet is typically accomplished using a simple optical reflection technique provided the angle is greater than the minimum angle where perforation occurs. More detailed ricochet can be accomplished as a function of the impact surface's harness and fragment size, shape and velocity.

## 4   Load

The loads on the structural elements, equipment and inhabitants are calculated from the modified blast and fragment environments.

## 4.1 Blast

The blast load is calculated as a function of the distance (range) from the detonation point to the component of interest. The load may be peak pressure for comparison to breach capacity or impulse for the windows. The impulse is calculated by integrating the impulse over the structural element.

## 4.2 Fragment

The fragment load is also calculated by integrating the impulse over the structural element. For equipment and personnel the load is typically momentum based.

# 5  Response

Most codes model structural response using pressure impulse techniques [8]. The damage to walls, beams, and columns is typically explicitly modelled. The loads determined from the time history approximations, modified for reflections and integrated to obtain impulse, are compared to the structural capacity of the various components of interest to determine damage. The damage is accumulated and used for evaluation of structural and personnel response. Typical structural response mechanisms are breach, shear and flexure failure. Only flexure failure is discussed here. Window damage and structural collapse will also be discussed.

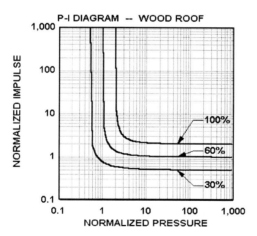

Figure 7:     Example pressure impulse diagrams.

## 5.1 Blast

A simple method for addressing flexural failure is to use threshold pressure-impulse diagrams. A pressure-impulse diagram (P-I diagram) for a given structural component is a plot of the combined values of the applied pressure and impulse that lead to a given level of structural damage. That is, a P-I diagram is a contour curve for a given damage level that is plotted as a function of the applied pressure and impulse. As an example, P-I diagrams for a wooden roof/floor are shown in fig. 7. In these charts, the applied pressure and impulse are normalized (scaled) with respect to structural and geometric properties of the structural component. P-I diagrams derived on the basis of structural dynamics principles and available test data have been obtained [6–8] for a large variety of structural components including reinforced concrete beams, columns and slabs, wood floors and roofs, steel beams and columns, reinforced and un-reinforced masonry walls (CMU), etc. The velocity of the structural debris can be derived by making the applied impulse equal to the momentum of the debris.

Figure 8:    Example of glass debris.

## 5.2 Windows

Glass breakage is also typically determined based on peak pressure and total impulse. See fig. 8 for typical glass breakage. Typical glass has an average stiffness of 0.8 psi/in, a maximum allowed deflection of 0.9 in, an average natural frequency of 0.2 rad/s and an average natural frequency period of 30 ms. Window failure occurs when the predicted dynamic deflection exceeds the maximum allowable dynamic deflection. Equations (6)–(8), shown below, are used to determine predicted dynamic deflection. First, the load duration is computed:

$$T_{dur} = 2 \times (Total\ impulse) / (peak\ pressure) \tag{6}$$

Next, the ratio of load duration to averaged natural period of the equivalent SDOF:

$$T_{ratio} = T_{dur} / T_{n\_avg} \tag{7}$$

The dynamic load factor (DLF) is calculated from $T_{ratio}$ and a curve fit equation. The dynamic load factor is used to predict the maximum deflection $\delta_{dyn}$ of the window by multiplying the DLF by the static deflection of the window.

$$\delta_{dyn} = (Peak\ pressure / average\ stiffness) \times DLF \qquad (8)$$

The static value is the deflection that would occur if the peak applied pressure were applied statically to the window. Breakage is assumed to occur if the maximum predicted deflection exceeds the maximum allowable deflection.

## 5.3 Collapse

In the load transmission approach, gravity loads are determined by "trickling" the weight of each unfailed component down a tree of supports; component failures are determined by comparing loads against capacities. See fig. 9 for an example of how the collapse model re-allocates the loads. In the matrix methodology, gravity loads are determined by assembling and solving a global stiffness problem (similar to a finite element method), and component failures are determined by comparing loads in each of the assumed response modes to capacities. The mass distribution in the structure changes when structural components and connections fail. It is generally assumed in the collapse methodologies that failed components and any equipment supported by them fall onto the components below them. The component on which they fall has an additional load to support.

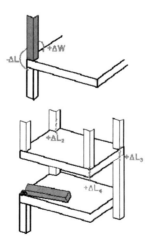

Figure 9:     Example of column collapse model.

## 6   Conclusion

Modelling of a terrorist attack in an urban environment requires a balanced understanding of the complex physical processes that occur and management of the inherent uncertainties associated with the modelling. This paper provided the key physics-based techniques required to accurately model a terrorist attack in an urban environment. Clearly there was insufficient space to cover all the nuances of the modelling. However, the blast and fragment environment that results from an IED detonation, the loads on an urban structure and the response of that structure, including progressive collapse, were covered in a manner to allow a basic understanding of the approach.

## References

[1]    Young, L.A., Streit B.K., Peterson, K.J., Read, D.L. & Maestas, F.A., *Effectiveness/Vulnerability Assessments in Three Dimensions (EVA-3D) Versions 4.1F and 4.1C User's Manual - Revision A.* Technical Report SL-TR-96-7000 prepared by Applied Research Associates, Inc. for U.S. Air Force Wright Laboratory, 1995.

[2]    York, A.R. & Harman, W., *Integrated Munitions Effects Assessment: A Weapons Effects and Collateral Effects Assessment Tool,* NBC Report, U.S. Army Nuclear and Chemical Agency, pp 30-37, Spring/Summer 2003.

[3]    Needham, C.E. & Crepeau, J.E., "The DNA Nuclear Blast Standard (1KT)," DNA 5648T, prepared by S-Cubed for the Defense Nuclear Agency, Alexandria, VA, 1981.

[4]    Kingery, C.N. & Bulmash, G., *Airblast Parameters from TNT Spherical Air Bust and Hemispherical Surface Burst,* Technical Report ARBRL-TR-02555, U.S. Army Armament Research and Development Center, Ballistic Research Laboratory, Aberdeen Proving Ground, MD, 1984.

[5]    Hacker, W.L. & Dunn, P.E., *Airblast Propagation and Damage Methodology,* Final Report, AMSAA Contract No. DAAA15-94-D-0005, Delivery Order 0013, Applied Research Associates, Inc. for U.S. Army Research Laboratory, AMSRL-SL-B, Aberdeen Proving Ground, MD, 1997.

[6]    Britt, J.R. & Little, C.D. Jr., *Airblast Attenuation Entranceways and Other Typical Components of Structures, Small-Scale Tests Data Report 1,* Technical Report SL-81V-22, U.S. Army Engineer Waterways Experiment Station, 1984.

[7]    Hikida, S. & Needham, C.E., *Low Altitude Multiple Burst (LAMB) Model, Volume I-Shock Description,* DNA 5683Z-1, prepared by S-Cubed for the Defense Nuclear Agency, Alexandria, VA, 1981.

[8]    *Facility and Component Explosive Damage Assessment Program (FACEDAP), Theory Manual, Version 1.2,* Contract No. DACA 45-91-D-0019, U.S. Army Corps Engineers, Omaha, NE, 1984.

# A study of the JWL equation of state parameters of dynamite for use in airblast models

B. J. Zapata & D. C. Weggel
*Department of Civil and Environmental Engineering,
UNC Charlotte, USA*

## Abstract

Two separate blast tests were conducted inside a conventional, unreinforced, brick building scheduled to be demolished. The small cylindrical explosive charges (less than 9 kg each), composed of dynamite sticks bundled together, were placed inside the building and detonated (in separate events) to study the blast resistance of the structure. The pressures generated by the blasts were recorded using a high speed data acquisition system. To better understand the complex pressure loading caused by the blasts for use in structural response modelling, the authors have undertaken a study to computationally model the explosive detonations. Advanced computational modelling is of interest because most tabular and other simplified blast load analysis techniques are inaccurate for the case of a close-in (but outside the detonation products) blast produced by a cylindrical charge. This paper presents the results of two dimensional airblast simulations performed using CTH, a shock physics hydrocode written by Sandia National Laboratories. The Jones-Wilkins-Lee (JWL) equation of state (EOS) coefficients for dynamite available in the literature are reviewed and then adjusted to reflect the properties of the dynamite used in these tests. CTH simulations are compared to the measured blast pressures and impulses to assess the ability of adjusted EOS parameters to model currently available commercial dynamite.

*Keywords: explosives, equation of state, airblast, bomb blast, impact and blast loading characteristics, interaction between computational and experimental results.*

# 1  Introduction

The Department of Civil and Environmental Engineering at the University of North Carolina at Charlotte conducted two blast tests inside a conventional, unreinforced, brick building before its scheduled demolition. The experimental program was designed to serve two purposes. The first purpose was to study the blast performance of a non-purpose-built, conventional brick building that, by definition, possessed representative dead loads, non-ideal boundary conditions, and the inherent characteristics resulting from the design and construction practices in the United States in the 1940s. The second purpose of the experiment was to study the capabilities of various analytical techniques to predict blast pressures for close-in, non-hemispherical (or non-spherical) charges.

This scenario is of interest because it results in non-planar blast waves as opposed to the planar blast waves generated by distant explosions. USACE [1] contains guidance for the analysis of blast loads generated by cylindrical charges, but this data is only tabulated for a few explosives relevant to military weaponry. Data has not been generated for commercial explosives such as the dynamite used in this work. Typically, defense laboratories or the Department of Energy in the United States are tasked with characterizing explosives to support ongoing weapons related programs. Within the government explosives community, the characterization of an explosive is typically project specific, and dynamite is not widely used for military applications because more powerful and stable explosives are readily available. While the use of precisely manufactured TNT would have simplified the airblast modeling effort, logistical considerations led the authors to use dynamite in this experimental program.

This article reports the results of an analytical and experimental program in which sticks of commercial dynamite (Unimax, Dyno Nobel, Salt Lake City, Utah) were bundled together and detonated in two separate blast events. First, JWL EOS coefficients available in the literature for commercial dynamites are reviewed. This data will be modified using a density scaling technique to develop JWL coefficients for a previously uncharacterized dynamite, Unimax. The adjusted JWL coefficients will be used with CTH, a shock physics hydrocode written by Sandia National Laboratories, to simulate the airblasts created by the dynamite charges. The simulations will be used to assess the adequacy of the new JWL coefficients for Unimax by comparing simulated pressures and impulses to those measured during the experiments.

# 2  Commercial dynamite in the United States

Dyno Nobel is the only manufacturer of nitroglycerin dynamites in North America today. In the 1980s there were still several nitroglycerin dynamite manufacturers as evidenced by Cooper [2]. Of the commercially manufactured dynamites, Unigel (made by Hercules) was widely considered the standard gelatin dynamite. Dyno Nobel acquired Hercules in 1985 and began manufacturing Unigel as its own product. With the rising use of bulk explosives like ANFO, the demand for dynamites decreased and Dyno Nobel became the sole manufacturer of nitroglycerin dynamites in North America.

Dyno Nobel currently manufactures two main nitroglycerin dynamite products, Unigel and Unimax. Their energetic and chemical properties are shown in Table 1. No test data was available regarding detonation pressure, so eqn (1) was used to compute it assuming the adiabatic gamma ($\Gamma$) was equal to 2.49.

$$P_{CJ} = \frac{\rho_o D_{CJ}{}^2}{(\Gamma+1)} \tag{1}$$

In this equation $\rho_o$ is the explosive's unreacted density and $D_{CJ}$ is the detonation velocity. The Unigel currently manufactured by Dyno Nobel is similar in density to that manufactured at the time of previous studies by other authors (see Table 2). In addition to Unigel, Dyno Nobel manufactures a more powerful dynamite called Unimax. Unimax is termed an extra gelatin dynamite by the manufacturer. The designation "extra" means that the composite explosive contains additional oxidizers. The "gelatin" designation refers to the nitroglycerin component which is combined with another agent to form a gel [2].

Table 1:     Properties of Dyno Nobel dynamites from LeVan [3].

| | Unimax | Unigel |
|---|---|---|
| **Detonation Velocity ($D_{CJ}$)** | 5856 m/s | 4300 m/s* |
| **Detonation Pressure ($P_{CJ}$)** | 14.7 GPa** | 6.89 GPa** |
| **Unreacted Density ($\rho_o$)** | 1.50 g/cc | 1.30 g/cc |
| **Relative Weight Strength** | 1.20 | 1.09 |
| **Nitroglycerin Ether Extract** | 26.2 % | 19.5 % |
| **Ammonium Nitrate** | 39.2 % | 67.0 % |
| **Sodium Nitrate** | 25.6 % | 7.40 % |
| **Heat of Explosion ($\approx$energy)** | 6.322 kJ/cc | 5.191 kJ/cc |

*Unigel's detonation velocity from Dyno Nobel proprietary computer code
**Values computed using eqn (1)

The primary quantities of interest when characterizing any explosive are the unreacted density and the two Chapman Jouguet (CJ) state parameters: detonation velocity ($D_{CJ}$) and detonation pressure ($P_{CJ}$). Table 2 shows that while Unigel's density has not varied considerably, there is a discrepancy between the manufacturer's detonation velocity (shown in Table 1) and that reported by other researchers. It should be noted that Unigel's detonation velocity in Table 1 was provided by the manufacturer as a minimum while Unimax's detonation velocity was experimentally determined by the manufacturer. This could partially explain the discrepancy between the two tables.

In addition to knowing the CJ state parameters for an explosive, an equation of state (EOS) is required to computationally model a blast. The Jones-Wilkins-Lee (JWL) EOS is one of the most commonly used due to its simplicity. The JWL EOS describes the adiabatic expansion of gaseous detonation products from

Table 2:    Summary of CJ parameters for dynamites from other researchers.

| Product Description | $\rho_0$ (g / cc) | $P_{CJ}$ (GPa) | $D_{CJ}$ (m / s) | Diameter (mm) | Ref. |
|---|---|---|---|---|---|
| Unigel | 1.26 | 12.8 | 5760 | Not Listed | [4] |
| Unigel | 1.294 | 12.0 | 5477 | Not Listed | [5] |
| Unigel | 1.262 | 12.0 | 5760 | Not Listed | [6] |
| Gelatin Dynamite | 1.50 | 15.9* | 6090 | 100 | [7] |
| Permissible Dynamite | 1.10 | 2.26* | 2680 | 45 | [7] |
| Ammonia Gelatin Dynamite | 1.50 | 15.4* | 5980 | 100 | [8] |
| Extra Dynamite | 1.36 | 6.55* | 4100 | 100 | [8] |

*Value computed using eqn (1)

the CJ state.  Although the JWL EOS is a mathematical abstraction of the thermochemical processes of detonation, it is sufficient for many engineering analyses and is easily implemented in hydrocodes.  The standard form of the JWL EOS is given by Lee et al. [9]

$$P(V,E) = A\left[1 - \frac{\omega}{R_1 V}\right]e^{-R_1 V} + B\left[1 - \frac{\omega}{R_2 V}\right]e^{-R_2 V} + \frac{\omega E}{V} \qquad (2)$$

In this equation, $P$ is the pressure and $A$, $B$, $R_1$, $R_2$, and $\omega$ are the JWL coefficients, $V$ is the relative volume which can be computed as $\rho_0/\rho$, where $\rho$ is the density of the detonation products, and $E$ is the energy.  The coefficients for this EOS are derived from tests in which a cylinder of explosive encased in copper is detonated at one end.  The cylinder wall velocity time history is recorded and compared to simulations, typically performed using LS-DYNA (LSTC, Livermore, CA).  The simulations are performed iteratively, adjusting the input JWL coefficients until the simulated cylinder wall velocity or displacement matches the test data.  Table 3 shows some of the JWL coefficients for Unigel dynamite found in the literature.  In the table, all variables are as defined above and $E_o$ is the explosive's available chemical energy.

It is worth briefly discussing the notion of ideality.  Penn et al. [6] define an ideal explosive as one in which there is a constant rate of energy release over a wide range of diameters while Souers et al. [10] defines an ideal explosive as one which follows Zel'dovich-von Neumann-Doring (ZND) theory and possesses a true CJ point under heavy confinement.  Powerful military explosives like PETN are ideal explosives.  Composite explosives like dynamite and ANFO typically do not display these characteristics.  Penn et al. [6] noted that, while ANFO's behavior was highly complex and would require a more complex EOS, dynamite could be approximated as an ideal explosive and thus the JWL EOS could be used.

Table 3: Summary of available JWL coefficients for dynamite in literature.

| | Unigel Penn *et al.* [6] | Unigel Hornberg [5] | Unigel Edwards *et al.* [4] |
|---|---|---|---|
| $\rho_o$ (g/cc) | 1.262 | 1.294 | 1.26 |
| $P_{CJ}$ (GPa) | 12.0 | 12.0 | 12.8 |
| $D_{CJ}$ (m / s) | 5760 | 5477 | 5760 |
| $E_o$ (kJ / cc) | 5.1 | 5.1 | 4.04 |
| $\Gamma$ | 2.49 | Not Reported | 2.49 |
| $A$ (GPa) | 190.7 | 121.831 | 109.70 |
| $B$ (GPa) | 7.58 | 1.857 | 7.58 |
| $R_1$ | 4.4 | 3.60150 | 4.4 |
| $R_2$ | 1.4 | 0.86185 | 1.4 |
| $\omega$ | 0.23 | 0.20 | 0.23 |
| $C$ (GPa) | 0.627 | 0.549 | Not Reported |

It is interesting to note that the energy (or heat of explosion) provided by Dyno Nobel for Unigel in Table 1 is similar to the $E_o$ value for two of the coefficient sets listed in Table 3. Penn *et al.* [6] explicitly state that their $E_o$ value was based on the heat of formation of the detonation products at the CJ state, but Hornberg [5] and Edwards *et al.* [4] do not provide a clear indication of how they arrived at their energy values. $E_o$ is important to the JWL formulation because it is used to make the energy of the JWL EOS consistent with the explosive's available energy and is directly correlated to the airblast output.

## 3 Density scaling JWL coefficients

The experimental program described in this paper made use of Unimax, for which no well developed EOS coefficients exist. One simple method of generating JWL coefficients for an uncharacterized explosive like Unimax is to use its density to scale the JWL coefficients of another similar explosive. There are, however, very few methods available for engineers to perform such a scaling procedure. For very small adjustments, one density based adjustment method, as presented in Souers *et al.* [10], may be used. Small density adjustments of this type are typically required when analyzing multiple shots of the same explosive during a test series. In Lee *et al.* [9], where the JWL EOS is first presented, the authors provide another method of scaling JWL coefficients based on density for changes on the order of 10%. While the density scaling used in this article is approximately 15%, this latter method was still employed to investigate its applicability. Although it would be possible to use advanced thermochemical equilibrium codes as another method of generating JWL coefficients, such tools are not generally available to the public.

The scaling procedure can be described as follows. The original $\Gamma$ value, the new unreacted density, and the new detonation velocity are used to compute the new CJ pressure. The original $E_o$ is linearly scaled by the ratio of new to original density to obtain the new $E_o$. The original values of $R_1$, $R_2$, $\Gamma$, and $\omega$ are used in conjunction with the new values of $E_o$ and $P_{CJ}$ to solve three simultaneous equations that relate pressure, relative volume, and energy. The results of the solution of this system of equations are new values of $A$, $B$, and $C$. The full procedure is well documented in Lee *et al.* [9].

The three sets of JWL coefficients from the literature listed in Table 3 were compared to the Unigel currently manufactured by Dyno Nobel (see Table 1). The first two have similar energies ($E_o$) while the third is significantly lower. The third set was therefore discarded. The second set of coefficients from Hornberg [5] had unusual values for $R_1$ and $R_2$, and this caused the adjustment method to produce a negative value for $B$, which was unacceptable. The second set was also discarded. The adjustment scheme described above was therefore directly applied to the JWL coefficients from Penn *et al.* [6]. While the manufacturer supplied heat of explosion and the JWL $E_o$ for Unigel compare favourably, there is no indication that Unimax's $E_o$ should also be closely related to the heat of explosion. Without any further knowledge, the original $E_o$ for Unigel was scaled based on the density ratio, rather than directly specifying the energy. The results of the JWL coefficient scaling procedure are provided in Table 4. This set of adjusted JWL parameters will be used to model the airblast from the experimental program.

Table 4:    Original and adjusted JWL coefficients.

| | Unigel Penn *et al.* [6] | Adjusted Unimax |
|---|---|---|
| $\rho_o$ (g/cc) | 1.262 | 1.50 |
| $P_{CJ}$ (GPa) | 12.0 | 14.7 |
| $D_{CJ}$ (m / s) | 5760 | 5856 |
| $E_o$ (kJ / cc) | 5.1 | 6.1 |
| $\Gamma$ | 2.49 | 2.49 |
| $A$ (GPa) | 190.7 | 234.4 |
| $B$ (GPa) | 7.58 | 9.51 |
| $R_1$ | 4.4 | 4.4 |
| $R_2$ | 1.4 | 1.4 |
| $\omega$ | 0.23 | 0.23 |
| $C$ (GPa) | 0.627 | 0.716 |

## 4   Blast tests

The dynamite charges were detonated in two separate rooms of an unreinforced brick masonry building. Figures 1 and 2 show plan views of Blast Chambers A and B, respectively. The drawings show the locations of the charge and sensors in each blast chamber; sensor elevations are measured relative to the finished floor of the chambers. In both locations, the chamber walls were instrumented with piezoelectric pressure transducers manufactured by PCB Piezotronics

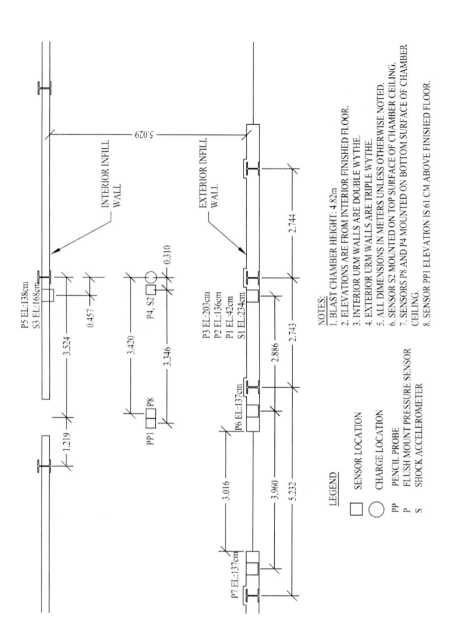

Figure 1:   Plan view of Blast Chamber A.

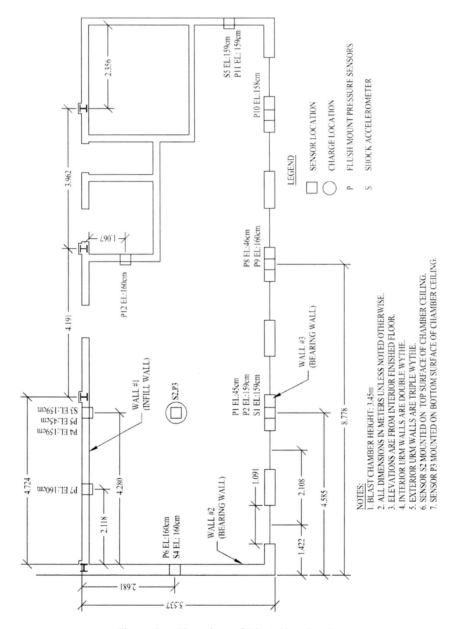

Figure 2:   Plan view of Blast Chamber B.

(Depew, New York).   The transducers were flush mounted on prefabricated metal plates mounted to the interior surface of the chamber walls.   The instruments were powered with 4 milliamps of ICP excitation and data recorded by National Instruments 4472 modules sampling at 100 kHz with 24 bit resolution.

The charge in Blast Chamber A was 6.53 kg and the charge in Blast Chamber B was 8.71 kg of Unimax sticks formed into upright cylinders. The charges were designed to damage, but not totally collapse the walls in each chamber. The height of both charges was approximately 41 cm. At detonation, the bottoms of the charges were at a height of 30 cm relative to the finished floor of the blast chamber. Both charges were detonated at their centers of mass with two instant electric blasting caps.

## 5   Airblast modelling

Airblast modelling was performed using CTH, a three dimensional shock physics hydrocode written and maintained by Sandia National Laboratories [11]. Simulations were performed using the built in JWL EOS. The detonation was modelled using the HEBURN feature of CTH. HEBURN allows the user to specify a detonation point and detonation velocity. The code then automatically handles the insertion of energy into the mesh to simulate detonation. In order to use the code's JWL EOS, the user supplies CTH with the constants $A$, $B$, $\omega$, $R_1$, $R_2$, $P_{CJ}$, $D_{CJ}$, and $\rho_o$. The code then computes all other necessary quantities automatically and performs a check to ensure that the specified JWL coefficients are consistent with the specified CJ state.

The simulation was performed in 2D axisymmetric cylindrical coordinates. Thus the explosive was placed at the center of the mesh, and a symmetric boundary was defined through the center of the charge. This symmetry combined with the use of adaptive mesh refinement (AMR) permitted highly resolved simulations. AMR was controlled using two refinement indicators, one which tracked high pressure gradients and the other which predicted the kinetic energy errors associated with unrefinement. The explosive material and the air shock were both zoned at 0.16 cm. A convergence study showed that this meshing scheme provided reasonably accurate pressures and very accurate impulses. The air used in the simulations was adjusted to account for atmospheric conditions at the time of the test. There was significant venting area in both blast chambers, so gas phase pressures were not observed in the experimental data. Interior shock reflections were however observed. In order to facilitate the 2D modelling effort, pressure and impulse comparisons were made for only the first reflection of the shock front. Furthermore, the slight angle of incidence (approximately 10 degrees) of the sensors in shot A relative to the charge will be ignored as its effect on the results is relatively insignificant.

## 6   Results

Airblast simulations of the two experiments were performed using the JWL parameters listed in Table 4 and the modelling techniques described in the previous section. Tables 5 and 6 compare experimental and predicted pressures and impulses in each blast chamber. Note that experimental and predicted impulses were computed as the time integrals of the pressure-time histories. The tables show that, in general, predicted and experimental impulses compared well. The average absolute impulse error was roughly 17 percent considering the comparisons in both blast chambers. Maximum reflected pressures, as expected, did not agree as well. On average, the error between experimental and predicted reflected pressures was 39 percent.

The highest observed impulse error was for sensor P1 in Blast Chamber B. Table 6 shows that CTH predicts almost 1.5 times the impulse recorded during the experiment. If one considers that charge B was almost 2.2 kg larger than charge A and the height of sensor P1 in each location was nearly the same, then it would be logical for the P1 sensor in shot B to record a higher impulse than in shot A. Looking at sensor P1 in Tables 5 and 6, however, this is clearly not the case. This would suggest an incomplete detonation of shot B or a malfunction of the P1 sensor for this shot. However, a definitive conclusion regarding the source of this error cannot be drawn from the experimental data alone.

Table 5:    Comparisons of experimental and predicted pressures and impulses from Blast Chamber A. Cell size of 0.16 cm.

| Sensor Location | Exp. Pressure (MPa) | Pred. Pressure (MPa) | Exp. Impulse (MPa ms) | Pred. Impulse (MPa ms) |
|---|---|---|---|---|
| P1 | 5.44 | 4.06 | 1.26 | 1.39 |
| P2 | 2.33 | 1.09 | 0.628 | 0.552 |
| P3 | 0.810 | 0.386 | 0.363 | 0.400 |
| P5 | 2.30 | 1.17 | 0.738 | 0.572 |

Table 6:    Comparisons of experimental and predicted pressures and impulses from Blast Chamber B. Cell size of 0.16 cm.

| Sensor Location | Exp. Pressure (MPa) | Pred. Pressure (MPa) | Exp. Impulse (MPa ms) | Pred. Impulse (MPa ms) |
|---|---|---|---|---|
| P1 | 11.0 | 7.79 | 1.26 | 1.90 |
| P2 | 1.50 | 1.21 | 0.593 | 0.614 |
| P4 | 1.85 | 1.21 | 0.627 | 0.614 |
| P6 | 1.06 | 0.579 | 0.655 | 0.496 |

## 7  Conclusion

The JWL coefficients for dynamite available in the literature were reviewed and the closest match to currently produced dynamites was selected. This set of JWL coefficients was used along with a density adjustment procedure to arrive at a set of JWL coefficients for a previously uncharacterized dynamite, Unimax. The adequacy of the JWL coefficients adjusted to model Unimax was examined by pressure and impulse comparisons to the experimental results of two blast tests. Tables 5 and 6 show that experimental and predicted impulses compared well, but reflected pressures did not.

Typically, maximum reflected pressures are more difficult to accurately simulate than reflected impulses since a much finer mesh is required to capture peak reflected pressures. However, a number of other factors could have contributed to the error in pressure comparisons. First, the JWL coefficients for Unimax were based on scaling Unigel's parameters. While Unigel and Unimax are both nitroglycerin dynamites, Table 1 shows that the makeup of their oxidizers (ammonium nitrate and sodium nitrate) varies considerably. The scaling procedure used in this work can only account for the effect of density and thus chemical makeup will not be taken into account unless separately considered. Another possible contributing factor is that the charges were bundled sticks of dynamite, rather than a large diameter monolithic charge. The behaviour of a bundled charge relative to the typical case of a monolithic charge has not been investigated.

Despite the pressure errors observed in these simulations, the use of CTH, coupled with the modified JWL coefficients for Unimax, provided reasonably accurate predictions of the impulses observed during the experimental program. Even though a more accurate JWL definition could better model blast pressures, the JWL coefficients presented in this work are sufficient for engineers to use in computing blast loads for mechanics analyses that are impulse dominated.

## References

[1]  United States Army Corps of Engineers (USACE), *Design and Analysis of Hardened Structures to Conventional Weapons Effects*, UFC 3-340-01.

[2]  Cooper, P.W., *Explosives Engineering*, Wiley: New York, 1996.

[3]  LeVan, B., Personal communication, January 2007, Dyno Nobel, Salt Lake City, Utah.

[4]  Edwards, C.L., Pearson, D.C. & Baker, D.F., *Draft Ground Motion Data from the Small-scale Explosive Experiments conducted at the Grefco Perlite Mine near Socorro, New Mexico.* Report No. LA-UR-94-1003, Los Alamos National Laboratory, 1994.

[5]  Hornberg, H., Determination of fume state parameters from expansion measurements of metal tubes. *Propellants, Explosives, and Pyrotechnics*, **11**(1), pp. 23-31, 1986.

[6]  Penn, L., Helm, F., Finger, M. & Lee, E., *Determination of Equation-of-State Parameters for Four Types of Explosive.* Report No. UCRL-51892, Lawrence Livermore Laboratory, 1975.

[7]  Souers, P.C., Vitello, P., Esen, S., Kruttschnitt, J. & Bilgin, H.A., The effects of containment on detonation velocity. *Propellants, Explosives, and Pyrotechnics*, **29**(1), pp.19-26, 2004.

[8]  Sadwin, L.D. & Junk, N.M., Lateral shock pressure measurements at an explosive column. *Fourth Symposium on Detonation*, US Naval Ordnance Laboratory: White Oak, Maryland, pp. 92-95, 1965.

[9]     Lee, E.L., Hornig, H.C. & Kury, J.W. *Adiabatic Expansion of High Explosive Detonation Products*. Report No. UCRL-50422, Lawrence Radiation Laboratory, 1968.

[10]    Souers, P.C., Wu, B. & Haselman, L.C. *Detonation Equation of State at LLNL*, 1995. Report No. UCRL-ID 119262 Rev 3, Lawrence Livermore National Laboratory, 1996.

[11]    McGlaun, J.M. & Thompson, S.L., CTH: A three-dimensional shock wave physics code. *International Journal of Impact Engineering*, **10(1-4)**, pp.351-360, 1990.

# A comparison of hydrodynamic and analytic predicted blast pressure profiles

G. M. Stunzenas & E. L. Baker
*US Army Armament Research Development and Engineering Center, USA*

## Abstract

Modeling the structural response to blast relies on accurate descriptions of the blast loading pressure profiles. Traditionally, empirically based blast pressure histories are used for this modeling. However, the structural response and geometric configuration can strongly affect the blast loading profile, particularly for close-in blast loading configurations. As a result, high rate continuum modeling is being increasingly applied to directly resolve both the blast profiles and structural response. A variety of computer models exist for the purpose of analyzing blast pressures associated with different types of explosive charges and ranges. These computer models range from simple empirically based analytic models based off of the Hopkinson cube root scaling to multi-physics high rate finite element approaches, commonly known as "hydrocodes", which are capable of tracking shocks through the conservation equations of continuum mechanics. The purpose of this paper is to provide comparisons of blast profiles predicted by analytic models with a hydrodynamic model at various standoffs. Three computer models; BlastX, Conwep, and ALE3D, were used to model the detonation of five pounds of TNT. Pressure profiles for various standoffs were gathered from each computer model and compared. The ALE3D result is greatly dependant on the mesh size and appears to converge to the BlastX and Conwep solutions with increased mesh resolution.
*Keywords: blast, explosives, modelling.*

## 1 Introduction

Modeling the structural response to blast relies on accurate descriptions of the blast loading pressure profiles. Traditionally, empirically based blast pressure histories are used for this modeling.  However, the structural response and geometric configuration can strongly affect the blast loading profile, particularly for close-in blast loading configurations [1–3].  As a result, high rate continuum modeling is being increasingly applied to directly resolve both the blast profiles and structural response. An explosion produces shock waves in air, which extend outward from the point of detonation. This shock wave is composed of a highly nonlinear shock front, which decays as the distance from the source increases. This nonlinearity is characterized by a sharp, instantaneous increase in pressure, called the peak incident overpressure. The velocity of the shock is supersonic in the medium in which it travels. The gas molecules behind the shock travel at a lower particle velocity and generally make up what is referred to as the shock wind. As the volume in which the shock travels increases, the peak pressures associated with the shock decrease [2]. A variety of computer models exist for the purpose of analyzing blast pressures associated with different types of explosive charges and ranges.  These computer models range from simple empirically based analytic models based off of the Hopkinson cube root scaling to multi-physics high rate finite element approaches, commonly known as "hydrocodes", which are capable of tracking shocks through the conservation equations of continuum mechanics. The purpose of this paper is to provide comparisons of blast wave profiles in open air predicted by analytic models with a high rate continuum model at various standoffs.  Open air blast profiles were predicted using two analytic models: Conwep, BlastX, and a high rate continuum model: ALE3D.  A comparison of the peak incident overpressure was made for a 5 pound spherical TNT charge at various distances from the charge.

## 2  Analytic blast modeling

Conwep blast calculations are conducted using a set of empirically based equations and curves.  These empirically based scaled blast curves have been generated for spherical and hemispherical TNT charges [2].  These curves are used to predict blast variables including time of arrival, impulse, peak incident and peak reflected pressures.  These curves were created based on Hopkinson cube root scaling [3], which relates the characteristic properties of the blast wave from an explosion of one energy level to that of another.  According to this scaling law, the pressure at a certain distance from the charge is proportional to the cube root of the energy yield. Figure 1 presents scaled blast relationship curves [2] and is representative of how Conwep computes peak blast pressures and other blast characteristics for an open air detonation using a spherical TNT charge.

A scaled distance is calculated, which is the ratio of the distance from the charge to the cube root of the explosive weight.  This scaled distance is then used to compute all of the variables associated with the particular blast wave of interest. Table 1 shows the peak incident overpressure computed by Conwep at various distances from the charge. This table also shows the results of using the chart alone with the scaled distance calculations. Based on the similar results,

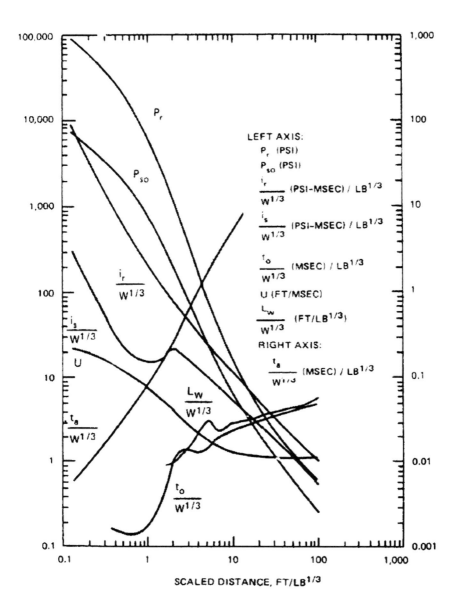

Figure 1: Blast wave characteristics vs. scaled distance for spherical TNT charge.

this demonstrates how Conwep uses this particular curve and scaling law to calculate air blast properties at various ranges.

One thing worth noting is that fig. 1 is on a log-log scale, which tends to makes accurate values difficult to assess from the graphical representations. This is likely the cause of any discrepancy between using the chart alone and the Conwep results. BlastX treats shock wave effects with a ray-based semi-empirical model. Similar to Conwep, it uses tabular blast data for spherical and cylindrical explosive charges. The blast data tables are based on hydrocode calculations for a 1 kg charge (of various explosives) at a standard set of atmospheric conditions. Results for other charge weights and atmospheric conditions are obtained similar to Conwep using Hopkinson scaling, as discussed above. BlastX uses the tabular values to calculate wave forms by interpolation of blast pressures, particle velocity, and density that were computed using the 1 kg spherical charge [5, 6].

Table 1:   Comparison between peak pressures obtained from Conwep and fig. 1.

| Distance from Charge (ft) | Conwep Pressure Results (PSI) | Using Chart and Scaled Distance (PSI) |
|---|---|---|
| 2.5 | 425.7 | 410 |
| 3 | 295.5 | 298 |
| 3.5 | 213.8 | 200 |
| 4 | 160 | 150 |
| 4.5 | 123.2 | 110 |
| 5 | 97.17 | 95 |
| 5.5 | 78.26 | 80 |
| 6 | 64.18 | 68 |
| 6.5 | 53.46 | 60 |

## 3   High rate continuum blast modeling

ALE3D is an arbitrary Lagrangian/Eulerian high rate finite difference hydrocode which is used to model fluid and solid elastic-plastic response of materials. A mesh is used to define a volume in space, and the conservation equations (mass, momentum, and energy) of continuum mechanics are applied and integrated through time, giving an updated nodal response to different forces, pressures, stresses, and strains [4]. The TNT charge was modeled using a standard Jones-Wilkins-Lee (JWL) detonation products equation of state. The air was modeled using a constant gamma equation of state.  Table 2 presents the explosive products and air equations of state parameters.

Table 2:    TNT JWL parameters and air constant gamma parameters used for the ALE-3D calculations.

|  | Density (g/cc) | Gamma | CJ Pressure (GPa) | Det Velocity (cm/microsecond) | EOS Coefficients |
|---|---|---|---|---|---|
| **TNT JWL Parameters** | 1.63 | 2.66 | 17.7 | .689 | A = 3.712 Mbar<br>B = .03231 Mbar<br>R₁ = 4.15<br>R₂ = .95<br>Omega = .30 |
| **Air Constant Gamma Parameters** | .000129 | 1.4 | N/A | N/A | N/A |

# 4  Results

Conwep, BlastX, and ALE3D were used to compute the peak pressure profiles of a 5 lb spherical charge of TNT, detonated in open air. The distance from the charge was varied, and the results are summarized in the Table 3. Fig. 2 shows some images that were captured from the ALE3D simulation for the open airblast. Certain parameters were modified in ALE3D, while holding the mesh resolution constant to examine the effect on the produced blast profiles and discrepancy between the different model outputs. The mesh resolution used for this study was approximately 0.12 cells/mm, resulting in 5.8 million total cells.

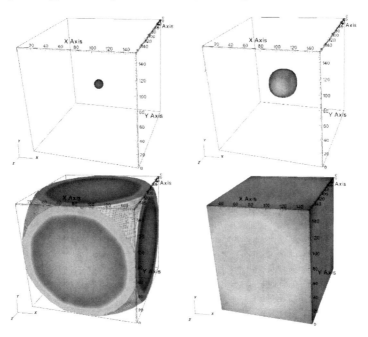

Figure 2:    ALE3D open airblast simulation images – blast front position.

   The initial purpose of this comparison was to study how the kinetic energy advection method affected the results. Since this term is quadratic in nature, rather than linear, energy can sometimes be lost if the velocity varies strongly in a calculation. The kinetic energy advection method can be set so that the energy lost computationally during integration for a shock is added to the internal energy (fracke = 1). There has been some debate about what this does to the resulting calculation, as the objective is to keep a strong shock without putting the material on a wrong adiabat [4]. ALE3D was run in Eulerian mode for the purposes of this comparison. Table 3 summarizes the results for BlastX, Conwep, and ALE3D at various distances with the kinetic energy advection term (fracke) set to 0 and 1. The information in Figure 3 depicts the results in this table:

Table 3:    Comparison of peak pressures for Conwep, BlastX, and ALE3D.

| Distance from charge (ft) | Conwep Pressure (psi) | BlastX Pressure (psi) | ALE3D - Eulerian (fracke = 1) Pressure (psi) | ALE3D - Eulerian (fracke = 0) Pressure (psi) |
|---|---|---|---|---|
| 2.5 | 425.7 | 400.3 | 226 | 220 |
| 3 | 295.5 | 278.4 | 175 | 171 |
| 3.5 | 213.8 | 212.8 | 142 | 144 |
| 4 | 160 | 161.5 | 119 | 121 |
| 4.5 | 123.2 | 123.6 | 99.4 | 103 |
| 5 | 97.17 | 96.57 | 90.2 | 86.1 |
| 5.5 | 78.26 | 76.78 | 82 | 82.6 |
| 6 | 64.18 | 62.7 | 73 | 77.5 |
| 6.5 | 53.46 | 52.25 | 69.7 | 69.9 |

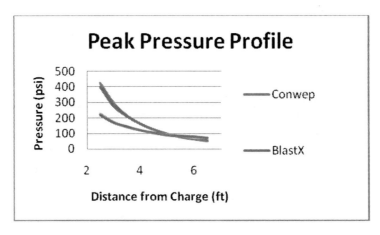

Figure 3:    Graphical comparison of peak pressures between Conwep, BlastX, and ALE3D at various standoffs.

It can be seen that the kinetic energy advection term did not have a significant effect on the calculations, but the results using ALE3D were much different than the empirically-based Conwep and BlastX. However, the Conwep and BlastX calculations produce similar blast results. In order to further investigate the differences between the high rate continuum modeling and the analytic blast models, several further modifications were made to the ALE3D computations. In an attempt to better track the shock wave propagation, monotonic artificial viscosity was used, rather than the default linear-quadrate rate dependant artificial viscosity. Subsequently, different mesh ratios were used to allow the shock to expand as it travelled away from the charge. This approach was investigated, as it is well known that large changes in mesh size affect shock propagation calculations. Finally, the resolution of the calculation was investigated, by increasing the number of elements to determine dependency on mesh size. Table 4 summarizes the results of the analysis at a distance of 2.5 feet from the charge. The results indicate that convergence to the analytic results appears to be occurring with increased mesh resolution.

Table 4:     Peak pressure results using ALE3D with modified input parameters.

| 2.5 Feet From Charge | Analysis Description | Resulting Pressure (psi) |
| --- | --- | --- |
| Run 1 | Eulerian, q weighting, slight increase in mesh size | 260 |
| Run 2 | q weighting Modequipotential rather than Eulerian | 262 |
| Run 3 | Eulerian, q weighting, different mesh ratios | 281 |
| Run 4 | Eulerian, q weighting, different mesh ratios | 284 |
| Run 5 | Eulerian, q weighting, finer mesh | 300 |
| Run 6 | Eulerian, q weighting, finer mesh | 326 |
| Run 7 | 10 million elements | 335 |
| Run 8 | 18 million elements | 339 |

## 5  Conclusion

More investigation is required if the true discrepancy between these computer models is to be determined, but based on the above analysis, it is quite evident that tracking shocks in ALE3D is greatly dependant on the mesh size. For the 5 LBS charge investigated, it appears that a mesh resolution of at least .15 cells/mm is required. The artificial viscosity term also had some impact on the calculations. Conwep and BlastX can not solely be depended upon when conducting blast analyses on structures simply due to the fact that they are empirical in nature. Much of the data calculated is based on curves and interpolated from existing databases. In the future, hydrodynamic codes, such as ALE3D will become increasingly necessary to solve these highly nonlinear and dynamic problems due to their capability of resolving both shock and structural response at the same time.

## References

[1]  Ngo, T., Mendis, P., Gupta, A. & Ramsay, J., Blast loading and blast effects on structures – an overview. *EJSE Special Issue: Loading on Structures*, pp. 76-91, 2007.

[2]  Headquarters, Department of the Army, Fundamentals of Protective Design for Conventional Weapons, *Technical Manual TM 5-855-1*, 3 November, 1986.

[3]  Headquarters, Department of the Army, Engineering Design Handbook, Explosions in Air, *AMC Pamphlet AMCP 706-181*, 15 July 1974.

[4]  Lawrence Livermore National Laboratories, *ALE3D High Performance Multi-Physics Simulations*, LLNL-MI-413853.

[5]  Commission on Engineering and Technical Systems (CETS), *Protecting Buildings from Bomb Damage: Transfer of Blast-Effects Mitigation Technologies from Military to Civilian Applications*, National Academy Press: Washington, D.C., 1995.

[6]  Ray J.C., Armstrong, B.J. & Slawson, T.R., Airblast environment beneath a bridge overpass, *Journal of the Transportation Research Board of the National Academies*, **1827**, pp. 63-68, 2003.

# Numerical determination of reflected blast pressure distribution on round columns

Y. Qasrawi[1], P. J. Heffernan[2] & A. Fam[1]
*[1]Queen's University, Canada*
*[2]Royal Military College of Canada, Canada*

## Abstract

Blast load parameters are reasonably easily determined for rectangular columns and can be derived from either the literature or numerous utility programs. Little compiled information is available with respect to exposed round columns. A series of numerical simulations were carried out to investigate the design pressure imparted to a round column by an explosion. The column diameter and charge weight were varied and pressure-time histories recorded at regular radial intervals along the face of the column from the closest point of first contact (front) to the extreme point at the side. A numerical model was created in AUTODYN, which simulated a blast wave diffracting around a rigid round cross section. The results indicate that as the diameter of the section increases, the peak reflected pressure at the point of first contact rapidly approaches that of a flat wall. However, the pressure varies sinusoidally between this peak at the point of first contact to a minimum equal to approximately the incident pressure at the furthest point at the side. The results support the obvious advantages when designing against blast to be realized by the use of round vs. rectangular columns, particularly when in the near field.
*Keywords: AUTODYN, numerical modelling, round column, reflected pressure, pressure distribution.*

## 1  Introduction

Fujikura *et al.* [1] have shown experimentally that the reflected over pressure experienced by a round column is substantially lower than the indicated design values. This fact and the intuitive expectation that round columns would deflect the blast wave indicate the need for modified design parameters. The first step in obtaining these modified parameters is determining the pressure distribution acting on a round column. In order to accomplish this, the problem was modelled numerically using the commercial software ANSYS AUTODYN. Both the size of the blast and the size of the column were varied to determine the effect of these parameters on the distribution. The model was verified against the current design values and it showed good agreement.

## 2   Numerical model

The problem was modelled using the commercial software ANSYS AUTODYN, an explicit analysis tool for modelling the nonlinear dynamics and interactions of solids, fluids, and gases.

The model was necessarily constructed in 3D as the cylindrical curvature of the columns could not be captured using 2D axial symmetry, while none of the blast energy dissipated in the third dimension if 2D planar symmetry were used.

The AUTODYN material library properties of air and TNT were used. Air was given an internal energy of 206.8 J, which results in the standard atmospheric pressure of 101.325 kPa.

The blast was initially modelled using 2D axisymmetry from the explosion out to 1.95 m (just before the blast wave interacts with the column) in a multi-

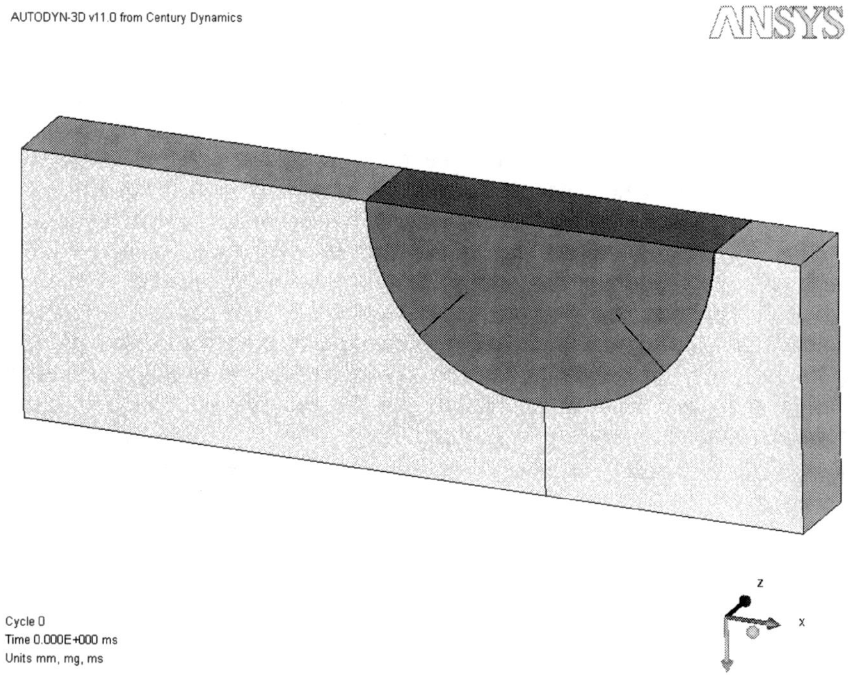

AUTODYN-3D v11.0 from Century Dynamics

ANSYS

Cycle 0
Time 0.000E+000 ms
Units mm, mg, ms

Figure 1:     Euler parts and 1000 mm radius column.

material Euler wedge. Then, AUTODYN's remapping capabilities were used to set the results of the 2D model as initial conditions for the 3D model in an ideal gas Euler mesh. The remapping and the use of quarter symmetry reduced the computation time and memory requirements considerably.

The 2D blast was modelled as a multi-material Euler wedge divided into 1950 elements and filled with air. The radius of the central TNT sphere was calculated using the predetermined TNT mass and a density of 1657 kg/m³.

The 3D model consisted of an Euler part surrounding a cylindrical Lagrange fill part. Fill parts are used to define the interaction boundaries in rigid coupling. The Euler gird was the same in all the models and only the column fill parts were varied. The dimensions of the Euler part were dictated by the range of the blasts (2000 mm), the largest column section (radius = 1000 mm), and the influence of the approximate outflow boundary condition ($\approx 500$ mm). Therefore the dimensions of the Euler part were $x = 4500$ mm, $y = 1500$ mm, and $z = 500$ mm. The Euler parts and a 1000 mm radius Lagrange fill part are shown in fig. 1.

The area of interest in the Euler part, where the blast interacted with the column, consisted of 10 mm cubic elements. Beyond the area of interest, AUTODYN's mesh grading capabilities were used to increase the element sizes geometrically by the maximum recommended rate of 1.2 in all three directions.

AUTODYN-3D v11 0 from Century Dynamics

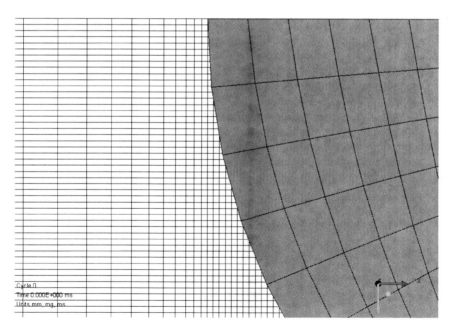

Cycle 0
Time 0.000E+000 ms
Units mm, mg, ms

Figure 2:    Grid detail at the interface between the 1000 mm radius column and the Euler part.

The column radii to be studied were chosen to be divisible by 10 mm to minimize the merging of Euler cells, which AUTODYN does automatically once a certain proportion of a cell is covered. AUTODYN also approximates circles using polygons, therefore, the number of elements in the cylindrical fill parts needed to be chosen carefully because they determined the precision of the circle. However, the restriction that the side of a Lagrange element must be larger than the smallest Euler element had to be adhered to too.

It was decided that all the cylindrical fill parts were to approximate Pi to two decimal places. Thus, the area of a polygon in terms of the number of its sides was equated to Pi to two decimal places multiplied by the radius squared. The number of sides was then solved. The number of sides used for all the columns was 56, which was obtained by having 15 cells across the radius in a type 2 cylindrical Lagrange part. This also gives a polygon side length of 11.2 mm which is larger than the minimum Euler element size of 10 mm. The interface between the 1000 mm radius column and the Euler grid is shown in fig. 2.

The default reflection boundary was applied to all planes of symmetry, and the approximate out flow boundary was applied to all surfaces where the blast is free to expand.

## 3  Investigation

The problem investigated was the variation of the reflected pressure (Pr) on the surface of a round column. The two variables that affect this distribution are the size of the column and the size of the blast. Therefore, in order to obtain a clear picture of the solution both the size of the column and the amount of explosive were varied to cover a practical range. The 100 mm radius lower bound for the column size was determined based on the fineness of the Euler grid in the model. While the 1000 mm radius upper bound was chosen as a practical limit. Thus, the column radii investigated were: 100 mm, 250 mm, 500 mm, 750 mm and 1000 mm with the two limiting cases of an unobstructed blast and a flat rigid wall.

For each of these seven cases, the blast was varied between a Z value of 0.8 m/kg $^{-1/3}$ to 2.4 m/kg $^{-1/3}$ in 0.4 m/kg $^{-1/3}$ increments resulting in 35 total runs. Z is the scaled distance and is given by the formula in eqn (1) below.

$$Z = \frac{R}{\sqrt[3]{W}}.$$

(1)

where $R$ is the range and $W$ is the weight of TNT.

To capture pressure measurements on the surface of the columns, nine gauges were placed radially at 11.25 degree increments around the circumference from the point closest to the blast to the point where the radius is perpendicular to the original point.

## 4  Verification

The model was verified against the design charts in TM 5-855 [2]. The values verified were the two limiting cases of the side-on pressure of an unobstructed blast and the pressure reflected off of an infinite flat rigid obstacle. The chart and model values are shown in table 1 below. The results showed very good agreement. The average per cent difference of the values was 12% for the side-on pressure and 21% for the reflected pressure.

## 5  Results and discussion

The surface plot in fig. 3 shows the variation of reflected pressure with respect to the radius of the column and the radial location around the circumference for a given Z value.

Table 1:    Verification of model pressure values from model and charts.

| Z $(m/kg^{-1/3})$ | Chart | | Model | |
| --- | --- | --- | --- | --- |
| | $P_{so}$ (MPa) | $P_r$ (MPa) | $P_{so}$ (MPa) | $P_r$ (MPa) |
| 0.8 | 1.75 | 10.0 | 1.5 | 8.9 |
| 1.2 | 0.7 | 3.5 | 0.67 | 2.9 |
| 1.6 | 0.4 | 1.5 | 0.37 | 1.15 |
| 2.0 | 0.2 | 0.6 | 0.21 | 0.44 |
| 2.4 | 0.15 | 0.3 | 0.197 | 0.38 |

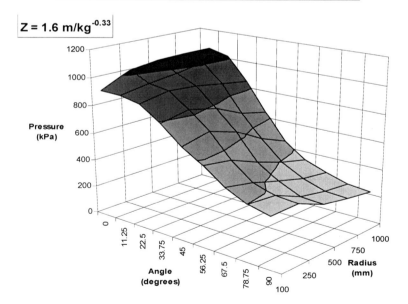

Figure 3:    Surface plot of variation of pressure with respect to radius and angle for a given Z.

The effect of the radius and radial location on the reflected pressure can be isolated.

Examining the variation of pressure with respect to radius at the point of first contact shows that the pressure reaches its maximum value rapidly as the radius is increased. It should be noted that if the radius axis started at zero, then the pressure value would be the side on pressure. The maximum pressure reached was approximately 90% of the reflected pressure of a rigid wall.

The cross sections of the plot at an angle of 0° for all the Z values studied are shown in fig. 4. The plots are a ratio of the reflected pressure and side on pressure for clarity, as the reflected pressure range would otherwise be too large to discern the plots for larger Z values. This plot shows that the trend of the initially rapidly increasing reflected pressure and the ceiling that is reached occurs for all the Z values studied. Also, that the asymptote is reached at a radius of about 250 mm in all cases.

The pressure variation with respect to radius at the edge of the column was found to be approximately equal to the side-on pressure at that location. It tended to be lower for larger diameters because the pressure wave had to travel the longer distance around the circumference whereas the free wave travelled in a straight line.

Similarly to the plot above, the ratio of the reflected pressure to the side on pressure is shown in fig. 5. Again the plots show that the reflected pressure remains approximately equal to the side on pressure at that location for all the Z values studied.

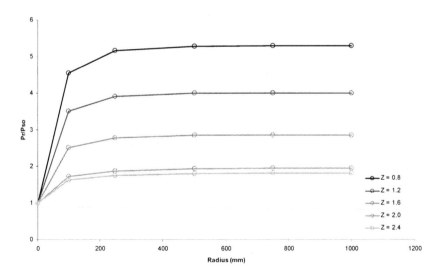

Figure 4:    Variation of the ratio of reflected pressure to the side-on pressure with respect to the radius.

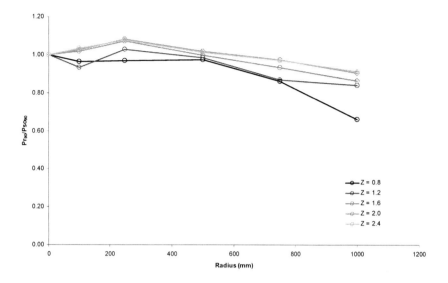

Figure 5:    Variation of reflected pressure over the side-on pressure with respect to the radius at the side of the column.

Between these two radial extremes, the pressure increases initially as the radius increases, but then starts to decrease as the radius continues to increase. This is because as the radius increases, the pressure wave must travel further to reach the radial location on the circumference, thus dissipating more energy. This is also more pronounced the further around the circumference the location of interest is.

The variation of pressure with respect to angle can be approximated using the sinusoidal curve fit represented by eqn (2). The result of this fit is shown in fig. 6.

$$P_{rc} = P_{so} + \left(P_r - P_{so}\right)\left(\frac{\cos 2\phi}{2} + \frac{1}{2}\right). \tag{2}$$

This formula was used to find the sinusoidal curve fit for all the column size and Z combinations. The fit showed very good agreement with the numerical results. However, it did over estimate the pressure for the larger columns. Again, this is attributed to the increased distance the wave travels and that the equation does not take distance into account.

Thus, one could obtain the reflected and side on pressure for a given blast from the standard sources, and then use a sinusoidal fit to approximate the pressure distribution around a circular column.

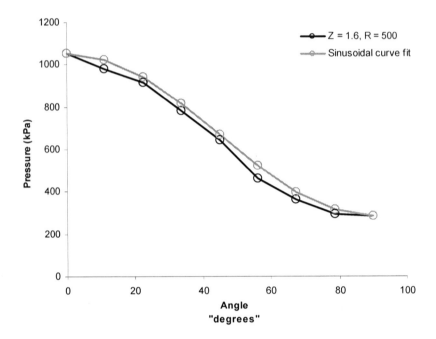

Figure 6:    Variation of reflected pressure with radial location with sinusoidal
             curve fit.

## 6    Future work

The established sinusoidal fit can be used to find an average or equivalent
pressure value to be used in design. A similar investigation needs to be
conducted for determining the distribution of impulse around a circular column
and finding an equivalent value that can be used in design. Perhaps most
importantly, the determined equivalent pressure and impulse values need to be
verified experimentally.

## 7    Conclusion

A numerical model was constructed in AUTODYN to investigate the pressure
distribution around a circular column. The model was verified and showed good
agreement with established values. It was found that as the column radius
increased the maximum reflected pressure at the point closest to the blast
approached a maximum of approximately 0.9 of the design value quickly. It was
also found that the pressure varied sinusoidally from this maximum to a
minimum at the side of the column approximately equal to the incident pressure.
A sinusoidal function was used to fit the distribution around the column with
good results and this curve fit can be used to find an equivalent design value.

# References

[1] Fujikura, S., Bruneau, M. & Lopez-Garcia, D., Experimental investigation of multihazard resistant bridge piers having concrete-filled steel tube under blast loading. *ASCE Journal of Bridge Engineering*, **13(6)**, pp. 586-594, 2008.

[2] US Department of the Army Technical Manual, TM 855. *Design and Analysis of Hardened Structures to Conventional Weapons Effects*, Washington, USA, 2002.

# Theory and calibration of JWL and JWLB thermodynamic equations of state

E. L. Baker[1], D. Murphy[1], L. I. Stiel[2] & E. Wrobel[1]
[1]US Army Armament Research Development and
Engineering Center, USA
[2]New York Polytechnic University, USA

## Abstract

Structure geometric configuration and response can be strongly coupled to blast loading particularly for close-in blast loading configurations. As a result, high rate continuum modeling is being increasingly applied to directly resolve both the blast profiles and structural response. In this modeling, the equation of state for the detonation products is the primary modeling description of the work output from the explosive that causes the subsequent air blast. The Jones-Wilkins-Lee (JWL) equation of state for detonation products is probably the currently most used equation of state for detonation and blast modeling. The Jones-Wilkins-Lee-Baker (JWLB) equation of state is an extension of the JWL equation of state that we commonly use. This paper provides a thermodynamic and mathematical background of the JWL and JWLB equations of state, as well as parameterization methodology. Two methods of parameter calibration have been used to date: empirical calibration to cylinder test data and formal optimization using JAGUAR thermo-chemical predictions. An analytic cylinder test model that uses JWL or JWLB equations of state has been developed, which provides excellent agreement with high rate continuum modeling. This analytic cylinder model is used either as part of the formal optimization or for post parameterization comparison to cylinder test data.
*Keywords: blast, explosives, equation of state, modelling.*

## 1   Introduction

Structure geometric configuration and response can be strongly coupled to blast loading particularly for close-in blast loading configurations. As a result, high rate continuum modeling is being increasingly applied to directly resolve both the blast profiles and structural response. Modeling the structural response to blast relies on accurate descriptions of the blast loading pressure profiles. When high rate continuum modeling is directly applied for the blast calculation, the explosive produced blast profile is calculated using detonation modeling of the high explosive event. In this modeling, the equation of state for the detonation products is the primary modeling description of the work output from the explosive that causes the subsequent air blast. The Jones-Wilkins-Lee (JWL) equation of state for detonation products is probably the currently most used equation of state for detonation and blast modeling. The Jones-Wilkins-Lee-Baker (JWLB) equation of state is an extension of the JWL equation of state that we commonly use. The purpose of this paper is to provide a thermodynamic and mathematical background of the JWL and JWLB equations of state, as well as parameterization methodology.

## 2   JWL equation of state

The JWL thermodynamic equation of state [1] was developed to provide an accurate description of high explosive products expansion work output and detonation Chapman-Jouguet state. For blast applications, it is vital that the total work output from the detonation state to high expansion of the detonation products be accurate for the production of appropriate blast energy. The JWL mathematical form is:

$$P = A\left(1 - \frac{\omega}{R_1 V *}\right)e^{-R_1 V *} + B\left(1 - \frac{\omega}{R_2 V *}\right)e^{-R_2 V *} + \frac{\omega E}{V *} \tag{1}$$

where $V*$ is the relative volume, $E$ is the product of the initial density and specific internal energy and $\omega$ is the Gruneisen parameter. The equation of state is based upon a first order expansion in energy of the principal isentrope. The JWL principal isentrope form is:

$$P_s \equiv Ae^{-R_1 V *} + Be^{-R_2 V *} + CV *^{-(\omega+1)} \tag{2}$$

For JWL, the Gruneisen parameter is defined to be a constant:

$$\omega \equiv \left.\frac{V * dP}{dE}\right|_{V*} \tag{3}$$

Energy along the principal isentrope is calculated through the isentropic identity:

$$d E_s = -P_s dV * \Rightarrow E_s = \frac{A}{R_1}e^{-R_1 V *} + \frac{B}{R_2}e^{-R_2 V *} + \frac{C}{\omega V *^\omega} \tag{4}$$

This relationship defines the internal energy referencing for consistency, so that the initial internal energy release is:

$$\Rightarrow E_0 = E_{CJ} - \frac{1}{2} P_{CJ} (V_0^* - V_{CJ}^*) \qquad (5)$$

The general equation of state is derived from the first order expansion in energy of the principal isentrope:

$$P = P_S + \frac{dP}{dE}\bigg|_{V^*} (E - E_S) = P_S + \frac{\omega}{V^*}(E - E_S) \qquad (6)$$

Combining eqns (2), (4), (6):

$$\Rightarrow P = A\left(1 - \frac{\omega}{R_1 V^*}\right) e^{-R_1 V^*} + B\left(1 - \frac{\omega}{R_2 V^*}\right) e^{-R_2 V^*} + \frac{\omega E}{V^*} \qquad (7)$$

From eqns (4) and (5) it can be seen the $E_0$ represents the total work output along the principal isentrope. For blast, this would represent the total available blast energy from the explosive.

## 3  JWLB equation of state

The JWLB thermodynamic equation of state [2] is an extension of the JWL equation of state. JWLB was developed to more accurately describe overdriven detonation, while maintaining an accurate description of high explosive products expansion work output and detonation Chapman-Jouguet state. The equation of state is more mathematically complex than the Jones-Wilkins-Lee equation of state, as it includes an increased number of parameters to describe the principal isentrope, as well as a Gruneisen parameter formulation that is a function of specific volume. The increased mathematical complexity of the JWLB high explosive equations of state provides increased accuracy for practical problems of interest. The JWLB mathematical form is:

$$P = \sum_n A_i \left(1 - \frac{\omega}{R_i V^*}\right) e^{-R_i V^*} + \frac{\lambda E}{V^*} \qquad (8)$$

$$\lambda = \sum_i \left( A_{\lambda i} V^* + B_{\lambda i} \right) e^{-R_{\lambda i} V^*} + \omega \qquad (9)$$

where $V^*$ is the relative volume, $E$ is the product of the initial density and specific internal energy and $\lambda$ is the Gruneisen parameter. The JWL equation of state may be viewed as a subset of the JWLB equation of state where two inverse exponentials are used to describe the principal isentrope ($n = 2$) and the Gruneisen parameter is taken to be a constant ($\lambda = \omega$).

## 4 Analytic cylinder model

An analytic cylinder test model that uses JWL or JWLB equations of state has been developed, which provides excellent agreement with high rate continuum modeling. Gurney formulation has often been used for high explosive material acceleration modeling [3], particularly for liner acceleration applications. The work of Taylor [4] provides a more fundamental methodology for modeling exploding cylinders, including axial flow effects by Reynolds hydraulic formulation. A modification of this method includes radial detonation product flow effects and cylinder thinning. The modifications were found to give better agreement with cylinder expansion finite element modeling [5]. One method of including radial flow effects is to assume spherical surfaces of constant thermodynamic properties and mass flow in the detonation products. The detonation products mass flow is assumed to be in a perpendicular direction to the spherical surfaces. A diagram of a products constant spherical surfaces cylinder expansion due to high explosive detonation is presented in fig. 1.

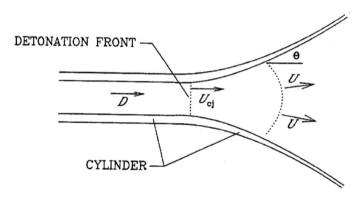

Figure 1: Analytic cylinder test model.

It should be noted that flow velocities are relative to the detonation velocity, $D$. If constant detonation product properties are assumed across spherical surfaces, the following model results using the JWLB thermodynamic equation of state.

Mass:

$$\rho_{cj} U_{cj} A_0 = \rho U A \tag{10}$$

Axial Momentum:

$$P_{cj} r_0^2 - P r^2 = \frac{m}{\pi} D^2 cos\Theta - \frac{m}{\pi} D^2 + \rho U^2 r^2 - \rho_{cj} U_{cj}^2 r_0^2 \; .... \tag{11}$$

Energy:

$$\rho_{cj} U_{cj} A_0 \left( \frac{U_{cj}^2}{2} + e_{cj} \right) + P_{cj} U_{cj} A_0 = \rho U A \left( \frac{U^2}{2} + e \right) + P U A \tag{12}$$

Principal Isentrope:

$$P = \Sigma_i A_i e^{\frac{-R_i \rho_0}{\rho}} + C \left(\frac{\rho_0}{\rho}\right)^{-(\omega+1)}, \quad de = -Pd\left(\frac{1}{\rho}\right) \qquad (13)$$

Taylor Angle:

$$v = 2D \sin \frac{\Theta}{2} \qquad (14)$$

Spherical Area:

$$A = \pi r^2 \frac{2(1-\cos\Theta)}{\sin^2 \Theta} \qquad (15)$$

The final equation set used for solution is:

$$(4) \Rightarrow P = \Sigma_i A_i e^{\frac{-R_i \rho_0}{\rho}} + C\left(\frac{\rho_0}{\rho}\right)^{-(\omega+1)} \qquad (16)$$

$$(4) \Rightarrow e_{cj} - e = \sum_i \frac{A_i}{\rho_0 R_i} \left( e^{\frac{-R_i \rho_0}{\rho_{cj}}} - e^{\frac{-R_i \rho_0}{\rho}} \right)$$
$$+ \frac{C}{\omega \rho_0} \left[ \left(\frac{\rho_0}{\rho_{cj}}\right)^{-\omega} - \left(\frac{\rho_0}{\rho}\right)^{-\omega} \right] \qquad (17)$$

$$(3) \Rightarrow \frac{U^2}{2} = \frac{U_{cj}^2}{2} + \frac{P_{cj}}{\rho_{cj}} - \frac{P}{\rho} + e_{cj} - e \qquad (9)$$

$$(2) \Rightarrow \frac{v^2}{2} = \left[ P\left(\frac{r}{r_0}\right)^2 - P_{cj} + \rho\left(\frac{r}{r_0}\right)^2 U^2 - \rho_{cj} U_{cj}^2 \right] \frac{C}{m\rho_0} \qquad (18)$$

$$(1),(5),(6) \Rightarrow \rho = \frac{\rho_{cj} U_{cj}}{U\left(\frac{r}{r_0}\right)^2} \left[ 1 - \left(\frac{v}{2D}\right)^2 \right] \qquad (19)$$

This set of equations is solved for a given area expansion, $(r/r_0)^2$ using Brent's method [6]. The spherical surface approach has been shown to be more accurate for smaller charge to mass ratios without any loss of agreement at larger charge to mass ratios. It should be recognized that this analytic modeling approach neglects initial acceleration due to shock processes [7] and is therefore anticipated to be more accurate as the initial shock process damps out. The model as expressed does not consider the fact that the cylinders thin during radial expansion. One simple way to account for this wall thinning is to assume that the wall cross sectional area remains constant and $r$ and $v$ represent the inside radius and inside surface wall velocity, respectively.

$$v_{out} = v \frac{r_{in}}{r_{out}} \; ; \; r_{out}^2 = r_{in}^2 + r_{out_0}^2 - r_{in_0}^2 \qquad (20)$$

## 5 Eigenvalue analytic cylinder model

High explosives are often aluminized for blast enhancement. Eigenvalue detonations are observed for some aluminized explosives [9]. For this reason, it was of interest to develop a modified analytic cylinder test model that provides a description of the detonation products isentropic expansion from the eigenvalue detonation weak point, rather than from the Chapman-Jouguet state. It was found that the most straight forward method of implementation of an eigenvalue detonation analytic cylinder model was to refit the isentrope associated with the eigenvalue weak point using eqn (13). In this way, equations 1-11 remain correct, except that eigenvalue weak point is used, rather than the Chapman-Jouguet state. With this approach, it is important to realize that the weak-point isentrope fit is not the same as the principal isentrope fit. The final form is:

$$P = \sum_i A_{wi} e^{\frac{-R_{wi}\rho_0}{\rho}} + C_w \left(\frac{\rho_0}{\rho}\right)^{-(\omega+1)} \tag{21}$$

$$e_w - e = \sum_i \frac{A_{wi}}{\rho_0 R_{wi}} \left( e^{\frac{-R_{wi}\rho_0}{\rho_w}} - e^{\frac{-R_{wi}\rho_0}{\rho}} \right)$$
$$+ \frac{C_w}{\omega \rho_0} \left[ \left(\frac{\rho_0}{\rho_w}\right)^{-\omega} - \left(\frac{\rho_0}{\rho}\right)^{-\omega} \right] \tag{22}$$

$$\frac{U^2}{2} = \frac{U_w^2}{2} + \frac{P_w}{\rho_w} - \frac{P}{\rho} + e_w - e \tag{23}$$

$$\frac{v^2}{2} = \left[ \begin{array}{c} P \left(\frac{r}{r_0}\right)^2 - P_w \\ + \rho \left(\frac{r}{r_0}\right)^2 U^2 - \rho_w U_w^2 \end{array} \right] \frac{c}{m\rho_0} \tag{24}$$

$$\rho = \frac{\rho_w U_w}{U \left(\frac{r}{r_0}\right)^2} \left[ 1 - \left(\frac{v}{2D_w}\right)^2 \right] \tag{25}$$

## 6 High rate continuum modeling comparison

ALE3D high rate continuum modeling (fig. 2) was compared to analytic cylinder test modeling using identical JWLB equations of state for TNT, LX-14 and PAX-30 for 1 inch diameter charges and 0.1 inch and 0.2 inch thick copper cylinders.

Figures 3, 4 and 5 present the comparison of the analytic cylinder test model to the ALE3D modeling for TNT, LX-14 and PAX-30 respectively. The analytic cylinder model slightly under predicts the velocities at 2 and 3 inside area expansions, but is in very close agreement by 6 and 7 inside area expansions. This is consistent with the fact that this analytic modeling approach neglects initial acceleration due to shock processes. Strong shock effects are typically observed in the 2 to 3 volume expansion region and are practically damped out by 6 volume expansions, where very close agreement between the analytic model and ALE3D results are observed.

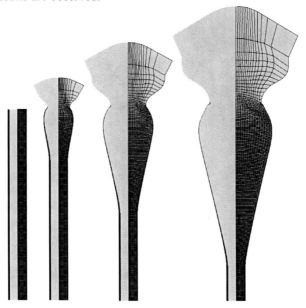

Figure 2:    Modeling at 10 μs intervals for 0.1″ thick copper cylinder.

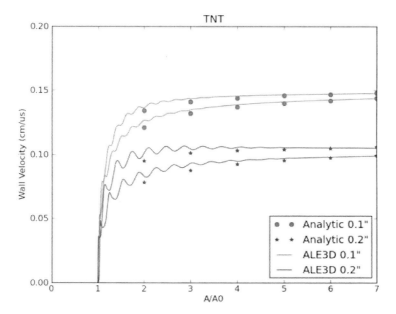

Figure 3:    TNT cylinder analytic model versus ALE3D.

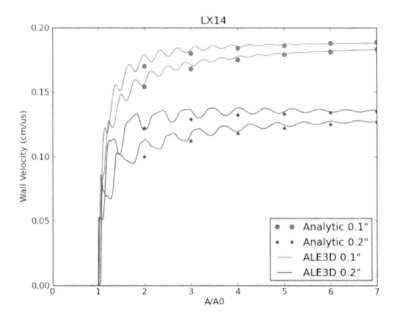

Figure 4:    LX-14 cylinder analytic model versus ALE3D.

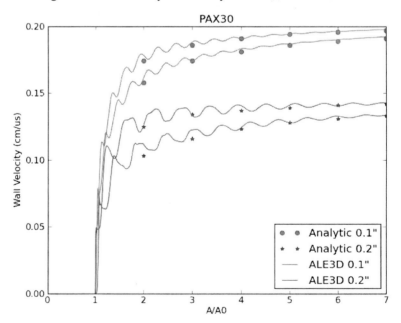

Figure 5:    PAX-30 cylinder analytic model versus ALE3D.

# 7 Parameterization

We use two methods of parameterization are used to calibrate the JWL and JWLB constants.    Both employ non-linear variable metric optimization techniques [2] for the parameterization process.    In the first method [2], the equation of state parameters are optimized to reproduce the experimental cylinder velocities using the analytic cylinder test model, as well as to reproduce a desired Chapman-Jouguet detonation velocity and pressure.   Typically, the total principal isentrope work output $E_0$ is also fixed to provide a desired total blast output.   The cylinder velocities are used in a cost function to be minimized, whereas the Chapman-Jouguet state and $E_0$ are treated as equality constraints. The second method of parameterization [8] is to directly fit the predicted pressure and Gruneisen parameter versus specific volume behavior predicted by the thermo-chemical equation of state computer program JAGUAR.   Formal non-linear optimization is used for the parameterization procedure.  The LX-14 high energy explosive example presented in Figure 4 used the technique of parameterization for the JWLB equation of state.   JWL and JWLB equation of states were parameterized for LX-14 using the JAGUAR predictions and non-linear optimization routines.   The resulting JWL and JWLB equations of state were then used to model a standard 1.2 inch outside diameter and 1 inch inside diameter copper cylinder test (0.1" thick wall) and compared to experimental data using the analytic cylinder test model. Table 1 presents the resulting outside cylinder velocity results at different inside cylinder cross sectional areas. The results clearly show the improved agreement to experimental data obtained when using the more mathematically complex JWLB mathematical form.    The

Table 1:    LX-14 JWL and JWLB cylinder test velocity predictions (Km/s) compared to experimental data.

| A/A0 | EXPERIMENTAL | ANALYTIC CYLINDER | |
| --- | --- | --- | --- |
| | | JWL | JWLB |
| 2 | 1.505 | 1.562 | 1.519 |
| 3 | 1.664 | 1.705 | 1.667 |
| 4 | 1.745 | 1.759 | 1.738 |
| 5 | 1.791 | 1.79 | 1.78 |
| 6 | 1.817 | 1.812 | 1.807 |
| 7 | 1.833 | 1.828 | 1.826 |
| | | % ERROR | |
| 2 | | 3.787 | 0.930 |
| 3 | | 2.464 | 0.180 |
| 4 | | 0.802 | 0.401 |
| 5 | | 0.056 | 0.614 |
| 6 | | 0.275 | 0.550 |
| 7 | | 0.273 | 0.382 |
| AVERAGED ERROR (%) | | 1.276 | 0.510 |

improved agreement is attributed to the improved agreement to the JAGUAR predicted detonation products behavior that is achieved using the JWLB form.

Similar to the LX-14, JWL and JWLB equation of states were also parameterized for PAX-30 using the JAGUAR predictions and non-linear optimization routines. The resulting JWL and JWLB equations of state were again used to model a standard 1.2 inch outside diameter and 1 inch inside diameter copper cylinder test (0.1″ thick wall) and compared to experimental data using the analytic cylinder test model. However, PAX-30 is an aluminized explosive that is known to produce eigenvalue detonations [9]. Table 2 presents the resulting outside cylinder velocity results at different inside cylinder cross sectional areas. Again, the results clearly show the improved agreement to experimental data obtained when using the more mathematically complex JWLB mathematical form. The results also show a slight improvement by using the eigenvalue analytic cylinder model that represents expansion from the weak point (w-point). Table 3 presents JWLB equation of state parameters for TNT, LX-14 and PAX-30, which were used in this study.

Table 2:    PAX-30 JWL cylinder test predictions compared to experiments.

| A/A0 | EXPERIMENTAL | JWL | JWLB | JWLB w-point |
|---|---|---|---|---|
| 2 | 1.499 | 1.599 | 1.55 | 1.541 |
| 3 | 1.682 | 1.759 | 1.702 | 1.703 |
| 4 | 1.774 | 1.823 | 1.780 | 1.779 |
| 5 | 1.827 | 1.862 | 1.831 | 1.825 |
| 6 | 1.859 | 1.89 | 1.868 | 1.856 |
| 7 | 1.883 | 1.911 | 1.897 | 1.879 |
| | | **% ERROR** | | |
| 2 | | 6.6711 | 3.4023 | 2.8019 |
| 3 | | 4.5779 | 1.1891 | 1.2485 |
| 4 | | 2.7621 | 0.3157 | 0.2818 |
| 5 | | 1.9157 | 0.2189 | 0.1095 |
| 6 | | 1.6676 | 0.4841 | 0.1614 |
| 7 | | 1.4870 | 0.7435 | 0.2124 |
| AVERAGED ERROR (%) | | **3.1802** | **1.0589** | **0.8026** |

# 8   Conclusions

An analytic cylinder test model has been developed by ARDEC for explosive equation of state calibration and verification. The analytic model was based on adiabatic expansion along the principal isentrope from the Chapman-Jouguet state. Additionally, an eigenvalue extended analytic cylinder expansion model has been developed based on isentropic expansion from the detonation

Table 3:    JWLB equation of state parameters for TNT, LX-14 and PAX-30.

| | TNT | PAX-30 | | LX-14 | |
|---|---|---|---|---|---|
| $\rho$ (g/cc) | 1.6300 | 1.885 | 1.909 | 1.820 | 1.8350 |
| E0 (Mbar) | 0.0657 | 0.13568 | 0.1376 | 0.102195 | 0.1032 |
| D (cm/$\mu$s) | 0.6817 | 0.8342* | 0.8429* | 0.86337 | 0.8691 |
| P (Mbar) | 0.1930 | 0.2419* | 0.2464* | 0.33529 | 0.3418 |
| A1 (Mbar) | 399.2140 | 406.224 | 405.3810 | 399.995 | 399.1910 |
| A2 (Mbar) | 56.2911 | 135.309 | 14.8887 | 20.1909 | 52.1951 |
| A3 (Mbar) | 0.8986 | 1.5312 | 1.49138 | 1.42441 | 1.59892 |
| A4 (Mbar) | 0.0092 | 0.006772 | 0.0076 | 0.02273 | 0.0249 |
| R1 | 28.0876 | 26.9788 | 13.2982 | 13.93720 | 27.4041 |
| R2 | 9.7325 | 10.6592 | 8.0204 | 7.230140 | 8.4331 |
| R3 | 2.5309 | 2.52342 | 2.4942 | 2.558910 | 2.6293 |
| R4 | 6.9817 | 0.335585 | 0.3566 | 0.736406 | 0.7498 |
| C (Mbar) | 0.0076544 | 0.013561 | .0135749 | 0.011016 | 0.385366 |
| $\omega$ | 0.345920 | 0.234742 | 0.234664 | 0.384733 | .0110204 |
| A$\lambda$1 | 58.2649 | 72.6781 | 66.6542 | 41.71970 | 68.6476 |
| A$\lambda$2 | 6.1981 | 5.64752 | 5.7776 | 6.83632 | 6.7497 |
| B$\lambda$1 | 2.9036 | 2.8728 | 3.1440 | 6.42909 | 4.1338 |
| B$\lambda$2 | -3.2455 | -3.10754 | -3.2552 | -4.47655 | -4.4607 |
| R$\lambda$1 | 25.5601 | 27.8109 | 25.5996 | 25.72540 | 26.2448 |
| R$\lambda$2 | 1.7034 | 1.71375 | 1.7099 | 1.71081 | 1.6977 |

* Eigenvalue weak point detonation state (not the Chapman-Jouguet state).

eigenvalue weak point, rather than from the Chapman-Jouguet state.  High explosives often include additive aluminium for blast effects.  This eigenvalue model is applicable to Al based explosives, such as PAX-30, that exhibit eigenvalue detonations.  The results for these explosives show only a very small reduction of explosive work output for eigenvalue detonations compared to Chapman-Jouguet detonations.  This is due to the fact that the Chapman-Jouguet principal isentrope and eigenvalue weak point isentrope lie very close to each other.  Excellent agreement between the analytic cylinder test and high rate continuum modeling predicted cylinder velocities is achieved when using the same JWL or JWLB parameters.

# References

[1]  Lee, E.L., Hornig, C. & Kury, J.W., *Adiabatic Expansion of High Explosive Detonation Products*, Lawrence Livermore Laboratory, Rept. UCRL-50422, 1968.
[2]  Baker, E.L., An application of variable metric nonlinear optimization to the parameterization of an extended thermodynamic equation of state. *Proceedings of the Tenth International Detonation Symposium*, eds. J.M. Short & D.G. Tasker, Boston, MA, pp. 394-400, 1993.
[3]  Gurney, R.W., *The Initial Velocities of Fragments from Bombs, Shells, and Grenades*, BRL Report 405, U.S. Army Ballistic Research Lab, 1943.

[4] Taylor, G.I., *Analysis of the Explosion of a Long Cylindrical Bomb Detonated at One End*, Scientific Papers of Sir G. I. Taylor, Vol 111:2770286, Cambridge University Press (1963), 1941.

[5] Baker, E.L., *Modeling and Optimization of Shaped Charge Liner Collapse and Jet Formation*, Picatinny Arsenal Technical Report ARAED-TR-92017, 1993.

[6] Brent, R., Algorithms *for Minimization without Derivatives.* Prentice-Hall: Englewood Cliffs, NJ, 1973. Reprinted by Dover Publications: Mineola, New York, 2002.

[7] Backofen, J.E., Modeling a material's instantaneous velocity during acceleration driven by a detonation's gas-push. *Proceedings of the Conference of the American Physical Society Topical Group on Shock Compression of Condensed Matter*, AIP Conf. Proc., Volume 845, pp. 936-939, 2006.

[8] Baker, E.L. & Stiel, L.I., Improved cylinder test agreement with JAGUAR optimized extended JCZ3 procedures. *Proceedings of the International Workshop on New Models and Numerical Codes for Shock Wave Processes in Condensed Media*, St. Catherine's College, Oxford, UK, 1997.

[9] Baker, E.L., Stiel, L.I., Capellos, C., Balas, W. & Pincay, J., Combined effects aluminized explosives. *Proceedings of the International Ballistics Symposium*, New Orleans, LA, USA, 2008.

# Prediction of airblast loads in complex environments using artificial neural networks

A. M. Remennikov[1] & P. A. Mendis[2]
[1]*School of Civil, Mining and Environmental Engineering, University of Wollongong, Australia*
[2]*Department of Civil and Environmental Engineering, The University of Melbourne, Australia*

## Abstract

Predicting non-ideal airblast loads is presently a complex computational art requiring many hours of high-performance computing to evaluate a single blast scenario. The goal of this research is to develop a method for predicting blast loads in a non-ideal environment in real time. The proposed method is incorporated in a fast-running model for rapid assessment of blast loads in complex configurations such as a dense urban environment or a blast environment behind a blast barrier. This paper is concerned with an accurate prediction of the blast loads from a bomb detonation using a neural network-based model. The approach is demonstrated in application to the problem of predicting the blast loads in city streets. To train and validate the neural networks, a database of blast effects was developed using the Computational Fluid Dynamics (CFD) blast simulations. The blast threat scenarios and the principal parameters describing the street configurations and the blast wall geometry were used as the training input data. The peak pressures and impulses were used as the outputs in the neural network configuration.
*Keywords: neural networks, blast loads, urban environment, explosion, numerical simulation.*

## 1  Introduction

Protecting civilian buildings from the threat of terrorist activities is one of the most critical challenges for structural engineers today. Events of the past few years have greatly heightened the awareness of structural designers of the threat of terrorist attacks using explosive devices. Extensive research into blast effects analysis and methods of protective design of buildings has been initiated in many countries to develop methods of protecting critical infrastructure and the built environment.

Although it is recognised that no civilian buildings can be designed to withstand any conceivable terrorist threat, it is possible to improve the performance of structural systems by better understanding the factors that contribute to a structure's blast resistance. One of such factors is the ability of the structural designer to accurately predict the blast loadings on structural components using analytical or numerical tools that take into account the complexity of the building, the presence of nearby structures, and the blast wave-structure interaction phenomena.

In recent years, the use of non-traditional tools based on artificial intelligence has received significant attention from the civil engineering researchers in relation to the systems that exhibit dynamic, multivariate and complex behaviours (e.g. wave forces, weather conditions, shock and impact problems). In this paper, an approach based on the neural network methodology is developed to train the neural networks capable of predicting the blast resultants in the complex geometries with reasonable accuracy, cost and minimum computing requirements. For brevity, the approach is demonstrated only in application to the problem of predicting the blast loads in city streets. The approach has also proven to be effective in other complex blast wave-structure interaction problems such as predicting the blast environment behind a rigid blast barrier.

## 2    Effect of adjacent structures on blast loads on buildings

Blast loads in simple geometries can be predicted using empirical or semi-empirical methods. These can be used to calculate blast wave parameters for hemispherical or spherical explosive charges detonated near the surface or in a free air to predict blast effects on isolated structures and structural components.

Events of the recent years have demonstrated that the most common source of unplanned explosions were terrorist devices in urban environment. In complex urban geometries, the blast wave behaviour can only be predicted from first principles using such numerical tools as AUTODYN [1], Air3D [2], and some others. Such tools solve the governing fluid dynamics equations and can be used to simulate three-dimensional blast wave propagation including multiple reflections, rarefaction and diffraction. In addition, Computational Fluid Dynamics (CFD) techniques can capture such key effects as blast focussing due to the level of confinement, shielding by other buildings and component failure (e.g. a window failure).

## 3    Neural networks

Artificial neural networks (ANNs) are computational models loosely inspired by the neuron architecture and operation of the human brain [3]. They are massively parallel; they can process not only clean but also noisy or incomplete data.

ANNs can be used for the mapping of input to output data without knowing 'a priori' a relationship between those data. ANNs can be applied in optimum design, classification and prediction problems.

An artificial neural network is an assembly (network) of a large number of highly connected processing units, the so-called nodes or neurons. The neurons are connected by unidirectional communication channels ("connections"). The strength of the connections between the nodes is represented by numerical values, which normally are called weights. Knowledge is stored in the form of a collection of weights. Each node has an activation value that is a function of the sum of inputs received from other nodes through the weighted connections.

The neural networks are capable of self-organisation and knowledge acquisition, i.e. learning. One of the characteristics of neural networks is the capability of producing correct, or nearly correct, outputs when presented with partially incomplete inputs. Further, neural networks are capable of performing an amount of generalization from the patterns on which they are trained. Most neural networks have some sort of "training" rule whereby the weight of connections is adjusted on the basis of presented patterns. Training consists of providing a set of known input-output pairs, patterns, to the network. The network iteratively adjusts the weights of each of the nodes so as to obtain the desired outputs within a requested level of accuracy. Error is defined as a measure of the difference between the computed pattern and the expected output pattern.

## 3.1 Multi-layer perceptron network (MLP)

The multi-layer perceptron (MLP) network trained by means of the back-propagating algorithm is currently given the most attention by application developers. The MLP network belongs to the class of layered feed-forward nets with supervised learning. A multi-layer neural network is made up of one or more hidden layers placed between the input and output layers as shown in Figure 1.

Each layer consists of a number of nodes connected in the structure of a layered network. The typical architecture is fully interconnected, i.e. each node in a lower level is connected to every node in the higher level. Output units cannot receive signals directly from the input layer. During the training phase activation flows are only allowed in one direction, a feed-forward process, from the input layer to the output layer through the hidden layers. The input vector feeds each of the first hidden layer nodes, the outputs of this layer feed into each of the second hidden layer nodes and so on.

At the start of the training process the weights of the connections are initialised by random values. During the training phase, representative examples of input-output patterns are presented to the network. Each presentation is followed by small adjustments of weights and thresholds if the computed output is not correct. If there is any systematic relationship between input and output and the training examples are representative of this, and if the network topology is properly chosen, then the trained network will often be able to generalize beyond learned examples. Generalization is a measure of how well the network performs on the actual problem once training is complete. It is usually tested by evaluating the performance of the network on new data outside the training set.

Generalization is most heavily influenced by three parameters: the number of data samples, the complexity of the underlying problem and the network architecture. Currently, there are no reliable rules for determining the capacity of a feed-forward multi-layer neural network. Generally, the capacity of a neural network is a function of the number of hidden layers, the number of processing units in each layer, and the pattern of connectivity between layers.

During first stage of creating an artificial neural network to model an input-output system is to establish the appropriate values of the connection weights and thresholds by using a learning algorithm. A learning algorithm is a systematic procedure for adjusting the weights and in the network to achieve a desired input – output relationship, i.e. supervised learning. The most popular and successful learning algorithm used to train multi-layer neural network is currently the back-propagation routine.

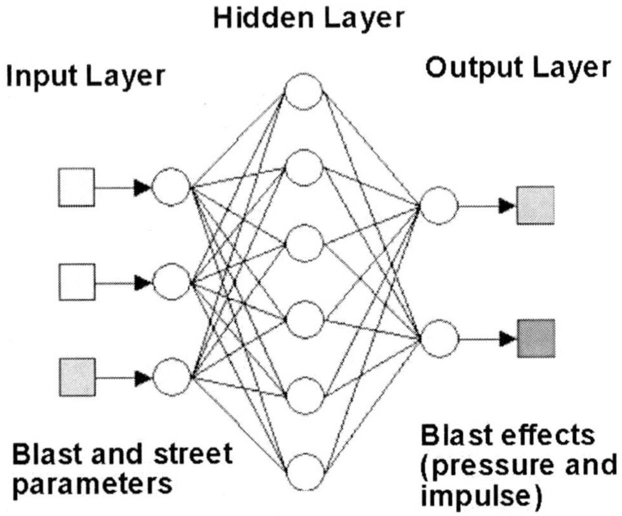

Figure 1: A simple back-propagating network for evaluation of blast effects in city streets.

### 3.2 Use of neural networks for predicting blast loads

The problem of blast in an urban environment is manifold; there are several different aspects that are of interest. Principally, among these are the direct effects of blast on people (both indoors and outdoors), the indirect effects of blast on people: from broken glazing, fallen masonry and collapsed building, and damage to buildings.

It is important to appreciate these problems if evacuation distances or safe areas within buildings are to be identified. All the above considerations have one common feature: they can only be quantified once the precise blast environment (in terms of pressure and impulse) is known throughout the region of interest.

Unfortunately, the effect of urban geometry on the propagation of blast waves is vast and complicated subject, and only recently has it begun to be approached in fundamental and systematic manner (Rose and Smith [4], Remennikov [5, 6]) and Remennikov and Rose [7]. In this paper, investigation of the effect of the street configurations on the pressure and impulse is one of the primary objectives. In order to quantify this effect, the pressure and impulse enhancement factors are defined as:

$$\text{Pressure enhancement factor } (E_p) = \frac{\text{Peak pressure}_{\text{street configuraton}}}{\text{Peak pressure}_{\text{unconfined burst}}}$$

$$\text{Impulse enhancement factor } (E_{imp}) = \frac{\text{Peak impulse}_{\text{street configuraton}}}{\text{Peak impulse}_{\text{unconfined burst}}}$$

One of the goals of the presented study was to develop a fast-running tool for predicting blast loads in an urban environment. This was accomplished by training an artificial neural network to approximate the pressure and impulse coefficients generated by a series of CFD numerical simulations for a selected street configuration and by varying the principal geometrical parameters of this street configuration. This possibility arises because every pressure monitoring location on the building façade can be identified by four independent parameters:

| Parameter | $[\text{m/kg}^{1/3}]$ |
|---|---|
| scaled street width | $w/W^{1/3}$ |
| scaled building height | $h/W^{1/3}$ |
| scaled distance along the street | $x/W^{1/3}$ |
| scaled height of the monitoring location above the ground | $y/W^{1/3}$ |

These four parameters can fully characterise the highly non-linear relationship between the explosive source weight, $W$, the standoff parameters $x$ and $y$, and the resulting peak pressure and peak scaled impulse. Therefore, these four parameters could be used as inputs to train the neural network using data generated by the CFD numerical simulations.

The problem of blast loads in an urban environment on the basis of CFD simulated numerical data is essentially a prediction (interpolation) problem. Since artificial neural networks are proving to be an effective tool for predicting values of blast loads, the basic idea in a neural network based approach is to train a network with patterns of the street and blast scenario parameters describing the spatial distribution of blast loads on building facades. This implies that each pattern represents the unique value of peak pressure and peak scaled impulse at each of the monitoring locations due to detonation of an explosive charge, $W$, at a particular location described by values of $x$ and $y$, in a street with principal parameters $w$ and $h$. Therefore, the patterns of the quantities describing the blast environment are used as inputs and the peak pressure and peak scaled impulse as outputs to train the neural network.

The training of a neural network with appropriate data containing the information about the cause and effect is a key requirement of a neural network approach. This means that the first step is to establish the training set, which can be used to train a network in a way that the network can predict the blast effects with the reasonable accuracy of 5 to 10% of the CFD-generated results. Ideally, the training set should contain data obtained by measurements, model tests or through numerical simulation, or through a combination of all three types of data.

In order to verify how well a trained network has learned the training cases, the trained network is tested by subjecting it to the training sets. The important generalization capability of a neural network for predicting blast wave parameters is tested by subjecting the trained network to data not included into the training sets (the so-called validating sets). How well a trained network is to generalize depends on the adequacy of the selected network architecture and the information on the blast load environment included in the training sets.

This study considers a broad range of scaled street widths: $w/W^{1/3} = 0.8$ to $3.2$ m/kg$^{1/3}$ and covers much of the range of practical interest. Similarly, scaled building heights from $h/W^{1/3} = 0.4$ to $2.4$ m/kg$^{1/3}$ is considered, which roughly equates to one to six storey buildings for a charge weight of 1000 kg. The scaled distance along the street was limited to the range 0.0 to 10.0 m/kg$^{1/3}$, which equates to a 100-m distance along the street for a 1000 kg explosive charge. Results of this study will lead to the possibility of using the information to predict blast loads (pressures and impulses) at any location in a street from knowledge of the charge weight, street width and building height.

### 3.3 Preparation of the dataset

The programme of numerical simulations was designed to cover most of the range of street widths that exist in real cities. Seven different street widths: $w = 8, 12, 16, 20, 24, 28$ and 32 m were considered. Similarly, the six different building heights: $h = 4, 8, 12, 16, 20$ and 24 m, which modelled one to six storey buildings, respectively, were used. For each street width, an additional simulation was performed for no buildings in the model. This last analysis produced side-on pressures and was used to evaluate values of the pressure and impulse enhancement factors. A total of 49 analyses were performed in this matrix to establish the enhancement factors data sets for training of the neural network.

A typical straight city street configuration is shown in Figure 2. The distance along the street $l$, the width $w$, and the location of a hypothetical hemispherical explosive charge $W$ are indicated. Figure 2 also shows the end view of the street where the building height $h$ and the building depth $d$ are indicated.

The parametric study was based on a 1000 kg TNT hemispherical charge, detonated on the centreline of the street. A computational domain $x = 125$ m (along the street) by $y = 48$ m wide and $z = 48$ m wide was used, requiring about

2,500,000 computational cells. Pressure monitoring points were located at 5 m intervals from $x = 0$ m to $x = 100$ m along the length of the street and in the vertical plane at levels $y = 2$ m, 6 m, 10 m, 14 m, 18 m and 22 m above the ground. The computational domain was extended sufficiently far in each direction to ensure that the presence of the boundaries did not affect the solutions at the pressure measuring locations.

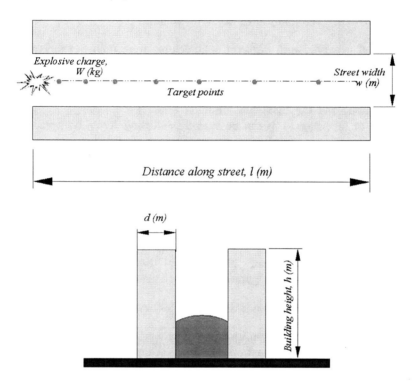

Figure 2: Plan and end view of a straight city street configuration.

### 3.4 Neural network implementation

The implementation stage of a neural network model typically includes the following two tasks: (1) data preparation, and (2) training and testing. Data preparation was conducted by performing a series of blast propagation analyses for the geometry shown in Figure 2 and using the test matrix of different building heights, street widths, and recording the values of the output parameters (peak pressures and peak scaled impulses) along the length of the street at the pressure measuring points. The developed dataset was collected and formatted to conform to the Air3D model input and output parameters identified at the design stage. Afterwards, the neural network was trained on the data and further tested on new datasets not used in training to ensure its generalisation performance.

NeuroShell 2 [8], Release 4.0, was the neural network software utilised in the present study to train the network. The program implements several types of neural networks and architectures, including backpropagation nets, multi-layer backpropagation nets, genetic adaptive general regression neural networks (GRNN), polynomial nets (GMDH), backpropagation nets with jump connections and some others.

**3.5 Performance and analysis of the ANN models**

Four neural networks with different architectures were trained. Processing the actual data through the trained neural network produces the network's predictions for each pattern in the data file. Statistical analysis of the ANN models is given in Table 1. Table 1 shows the $R^2$ (the coefficient of multiple determination) and the maximum absolute error (the maximum of |actual – predicted| of all patterns). It can be seen that all three tested backpropagation networks demonstrated fairly good accuracy of the predicted overpressures and impulses with the coefficients of multiple determination being very close to 1. The general regression neural network (GRNN) produced relatively low value of $R^2$ (0.794) for the predicted overpressures compared to the backpropagation networks (0.993). The same tendency is observed with regard to the maximum absolute error introduced by the selected networks. The maximum absolute errors for both the overpressures and impulses predicted by GRNN are indicative of higher errors and lower accuracy of the results. It can be concluded from the analysis of the performance of the trial networks that the 4-layer backpropagation network configuration produced the most accurate predictions.

Table 1:     Configuration and statistical performance of trial networks.

| No | Network Architecture | Configuration of networks | | | $R^2$ | |
|---|---|---|---|---|---|---|
| | | Input units | Hidden units | Output units | Pressure | Impulse |
| 1 | 3-layer backpropagation | 3 | 28 | 2 | 0.975 | 0.982 |
| 2 | 4-layer backpropagation | 3 | 14 /14 | 2 | 0.993 | 0.995 |
| 3 | 5-layer backpropagation | 3 | 8 / 8 / 8 | 2 | 0.993 | 0.995 |
| 4 | GRNN (general regression NN) | 3 | 1892 | 2 | 0.794 | 0.967 |

# 4    Results and discussion

As discussed earlier, the database of blast effects was generated using the blast propagation code Air3D for a typical straight street configuration. Several trial neural networks were tested and the best network with least absolute error in both output variables was selected. The 4-layer backpropagation neural network with 2 hidden layers with eight units in each hidden layer was found to produce the best predictions for the blast effects across all ranges of the input parameters.

The trained neural network was used to estimate the blast peak overpressure and peak impulse for a range of input parameters used to generate the database of blast effects from a series of CFD numerical simulations. Figure 3 demonstrates the comparison of the peak pressures and impulses along a straight street predicted by the two methodologies for the value of the scaled street width representing narrow city streets. From Figure 3, it can be seen that the blast effects predicted by the trained neural network show very good agreement with those obtained from the CFD simulations. In particular, the positive impulses predicted by the trained neural network are in very close agreement with the CFD generated values. Based on the obtained results, it has been proven that a neural network can be used as an effective tool for rapid prediction of blast effects in urban environments.

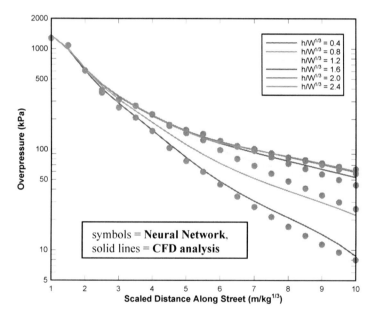

Figure 3: Comparison of CFD- and neural network-based blast overpressures and impulses along the street (scaled street width = 0.8 m/kg$^{1/3}$).

## 5 Conclusions

The main objective of this study was to evaluate a new approach of using artificial neural network for predicting the blast loads in complex city environments. A database of blast effects was built from a series of CFD blast simulations for a selected street configuration. This database was used to train and test the neural network. Analysis of the network's performance has demonstrated that the neural network can replace the time-consuming CFD analyses for a given street configuration and within the boundaries of the existing database of blast effects for this street configuration. It is expected to utilise the developed neural network-based technique as a part of an expert system that would be capable of predicting information about the likely injury and damage levels should an explosion occurs in a variety of urban environments based on the fast running predictive models.

## References

[1]    Century Dynamics, AUTODYN-2D & 3D Version 6.0 User Documentation, 2005.
[2]    Rose, T.A., Air3D User's Guide. 7.0: RMCS, Cranfield University, UK; 2003.
[3]    Dayhoff, J.E., *Neural network architectures: an introduction.* Van Nostrand Reinhold: New York, 1990.
[4]    Rose, T.A. & Smith, P.D., Influence of the principal geometrical parameters of straight city streets on positive and negative phase blast wave impulses. *International Journal of Impact Engineering*, **27**, pp. 359-376, 2002.
[5]    Remennikov, A.M., Evaluation of blast loads on buildings in urban environment. *Proc. of the 8th International Conference on Structures Under Shock and Impact*, Greece, pp. 73-82, 2004.
[6]    Remennikov, A.M., Blast effects analysis for commercial buildings in urban setting. *Proc. of the 18th Australasian Conference on the Mechanics of Structures and Materials (ACMSM)*, Perth, pp. 969 – 974, 2004.
[7]    Remennikov, A.M. & Rose, T.A., Modelling blast loads on buildings in complex city geometries. *International Journal of Computers and Structures*, **83**, pp. 2197-2205, 2005.
[8]    NeuroShell 2, Release 4.0, 2000. User's manual. Ward Systems Group Inc., Frederick, MD.

# Laboratory scale tests for internal blast loading

S. Kevorkian, N. Duriez & O. Loiseau
*Institut de Radioprotection et Sûreté Nucléaire, France*

## Abstract

The definition of blast loads applied to a structure of complex geometry is still nowadays a hard task when numerical simulation is used, essentially because of the different scales involved. As a matter of fact, modelling the detonation of a charge and its resulting load on a structure requires one to model the charge itself, the structure and the surrounding air, which rapidly leads to large size models on which parametrical studies may become unaffordable. Thus, on the basis of Crank-Hopkinson's law, an experimental set-up has been developed to support reduced scale structures as well as reduced scale detonating solid charges. As a final objective, the set-up must be used to produce the entry data for numerical assessments of the structural resistance. The set-up is composed of two mock-ups equipped with sensors and has been designed to conduct non destructive studies. In the context of security, the general aim is to study the effects of detonation shock waves inside the test installation and to test the influence of various openings. This set-up offers the possibility of measuring the loading in terms of pressure-time curves. The present paper summarizes the campaign of experiments performed in the year 2009 and gives the main features of the mock-up, the instrumentation and the pyrotechnics. During the campaign, internal blast tests have been conducted. Profiles of pressure versus time history are presented, taking into account relative positions of the explosive charge versus the gauges. The results obtained allow one to check that the Crank-Hopkinson's law is verified and shows the gas pressure influence.
*Keywords: blast waves, detonation, pressure measurements, reflections, gas pressure, safety.*

## 1  Introduction

Although important developments have taken place during the last decade, the definition of blast loads applied to a structure of complex geometry is still nowadays a hard task when numerical simulation is used, essentially because of the different scales involved (both in space and time). As a matter of fact, modelling the detonation of a charge and its resulting load on a structure requires modelling the charge itself, the structure and the air surrounding the charge and the structure, which rapidly leads to large size models on which parametrical studies may become unaffordable.

Since full-scale testing of realistic target geometries and realistic effects of charge position are often prohibitively expensive and time consuming, as far as detonation is involved, small-scale testing is a well proven means to assess blast loading. The most widely used method of blast scaling is Hopkinson's "cube-root" law for scaled distance, time and impulse.

This method has been used by IRSN to assess the pressure evolution in space in time though a free-field campaign of measurements [1]. In order to assess the pressure evolution due to small-scale detonations, IRSN realized a new campaign concerning the internal blast, which is described in this paper. Various authors used the method of blast scaling, in order to assess internal detonations. For instance, Miura *et al.* [2] showed a holographic interferometry system that permits the visualization and the measures of the propagation of an explosion of 10 mg of silver azide cylinder inside a small-scale closed room. Reference [3] presents results of explosions of a cylindrical charge made of composition B explosive inside several small-scale 3 and 4-wall cubicles of different sizes and shapes, these tests were made to establish a method and criteria for blast effects prediction.

Always relating to internal explosion studies, reference [5] presents the experimental measurements of pressure due to detonation of an explosive gaseous mixture (1 g eq TNT) confined in a hemispherical soap bubble inside an unvented small-scale structure. For solid explosives, reference [4] presents a comparison of an experimental explosion of 1 lb of C4 in a rectangular bunker and numerical calculations (Method of Images). Reference [6] shows numerical simulations done in order to study the influence of the building geometry, positions of explosion vent and ignition point. All these studies confirm the interest of studying explosion effects at small scale, mainly allowing one to capture a better understanding of the phenomena involved.

The experimental set-up described in the present paper is a laboratory scale set-up, constituted by a mock-up, able to bear the effects of detonations of solid explosives up to 16 g of TNT equivalent. In the context of security, the general aim is to study the effects of blast waves inside test installations and the influence of the openings.

## 2 Experimental set-up

The experimental set-up is composed of two mock-ups and sensors; it has been designed to conduct non destructive studies. The experimental campaigns are performed at the SNPE's Research Centre located at Le Bouchet (Vert-le-Petit, France). SNPE ensures all the pyrotechnics handling aspects of the experiments and also provide a spare data recording system.

## 2.1 Mock-ups

For this campaign, IRSN used two mock-ups made of steel, representing a parallelepipedic room.

In order to study the scale factor in a small-scale configuration for internal blast, two mock-ups with homothetic dimensions were made. The largest mock-up is 40 cm wide, 80 cm large and 40 cm high. The scale factor between the two different mock-ups was determined in such a way that a factor of four between the masses of explosives used in the two mock-up sizes was employed (i.e. the scale factor $\lambda$ is equal to 0.63 for the length).

For each size, three configurations are available; the first one is the mock-up without the front face, the second one with the front face with one opening and the third one with a full front face.

Threaded holes were made through the walls, soil and roof in order to allow insertion of air blast pressure gauges. Charges are supposed to be placed at the geometrical centre of the mock-up thanks to a guiding cap inserted from the top of the box.

## 2.2 Gauges

Eleven piezoelectric pressure transducers (Kistler, reference: 603B, range: 0–200 bars) are mounted on the mock-up. In the absence of pressure gauges, the holes in the walls are filled with a specific screw, so that they do not constitute unexpected venting or opening surfaces. Each pressure transducer is statically calibrated prior to the test campaign. The transducers are connected to an amplifier which is connected to the data acquisition system with electric microdot cables.

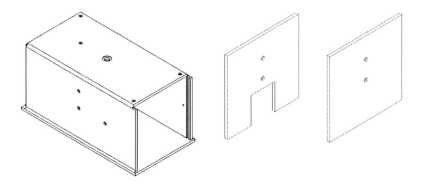

Figure 1:    The mock-up, the front faces with and without openings.

## 2.3 Detonating charges

A cylindrical charge of hexocire is initiated from one of its extremity using an electrical detonator. The masses used for the campaign described in the present paper are 1, 2, 3 and 4 g of TNT equivalent for the smallest mock-up and 4, 8, 12 and 16 g of TNT equivalent for the largest one.

## 2.4 Acquisition system

A LTT-186 data acquisition and transient recorder system connected to a PC has been used for the data acquisition in parallel with a NICOLET Odyssey.

# 3    Physical phenomena

The pressures observed after the explosion of a charge in a confined space is composed of two distinct phases.

The first phase is the reflected blast load impulse and the second one is due to the pressure of the gases created during the explosion [7, 8]. These phases are illustrated by fig. 2.

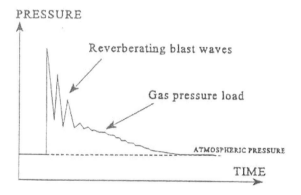

Figure 2:    Profile of pressure vs. time history.

## 3.1 Reflections

During an explosion, the shock wave is expanding in air up to the walls of the room, and then it is reflected (fig. 3).

For these experiments, an a priori estimate of the peak pressure was obtained using the software SHOCK [9].

In the literature, the maxima of pressure on a structure are most commonly estimated from scaled blast data or theoretical analyses of normal blast wave reflection from a rigid wall [8]. The subsequent shocks due to reflections are supposed to be attenuated.

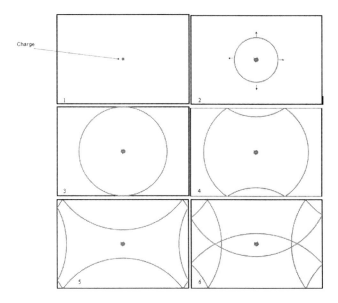

Figure 3:    Air blast propagation inside a rectangular room.

## 3.2  Gas pressure

When an explosion is produced, the detonation gas generated induces an increase of pressure inside the room. This pressure decreases more or less quickly depending of the number and the size of openings in the room. This phenomenon occurs after the propagation of the air blast.

The pressure and impulse due to the detonation gases have been estimated by using the software FRANG [10].

# 4 Internal blast campaign

## 4.1  Objectives

The aim of this campaign is to validate the reduced scale experimental concept for an internal blast.

The records of pressure vs. time history are used in order to:

- Check the scaling law or Crank-Hopkinson's law in an internal blast configuration;
- Determine the characteristic parameters of the blast waves – pressure, pulse, time of arrival – identified by the time evolution of pressure and compare the values obtained with values from abacuses available in the literature or out of calculation codes [9, 10];
- Observe experimentally the different physical phenomena composing the air blast : shock wave with reflections and gas pressure;

- Verify the trials reproducibility and the symmetry of measurements from symmetrically placed gauges;
- Show the difference of air blast pressure considering the position of the gauges in the room.

## 4.2 Description of the trials

For each size of the two mock-ups, four series of trials have been performed with four different masses of explosive.

For each masse of explosive, trials were performed with the three different types of opening.

For each type of opening, trials were conducted with two sets of positions for the gauges.

As a whole, 48 trials were performed, each one with eleven gauges in place. All the recordings of pressure vs. time history have been studied. Only the most representative are included in this article.

At first, the charge was placed at the exact geometrical centre of the mock-up, at the same height as the half-height gauges. By doing so, fragments of the detonator envelope came and hit the gauges after the charge initiation, causing several damages. To avoid this inconvenient, the charge was positioned at a height equal to one quarter of the overall mock-up height.

## 4.3 Position of charge and gauges

The positions of the gauges were chosen in order to verify:
- The reproducibility of the trials – gauges placed at the centre of the faces are the same for the two sets of gauges positions (K4 and K6, K9 and K11) and two gauges of the floor are the same (K1 and K3);

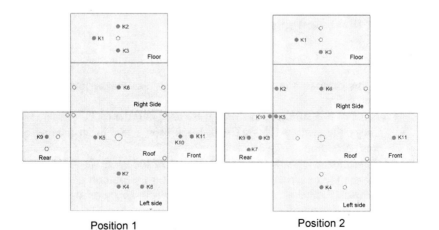

Position 1                Position 2

Figure 4:    Positions of the gauges for the mock-up with front face.

- The symmetry of the propagation of the air blast – considering the different axes, pairs of symmetrical gauges were studied (K4 and K6, K9 and K11, K2 and K3, K1 and K5, for the position 1);
- The profile of pressure for the largest number of geometrical positions inside the box:
  o centre of faces, nearest from the charge (K4 and K6), farthest from the charge (K9 and K11)
  o corners (K5 and K10 for the position 2)
  o edges (K2 for the position 2)
  o intermediate positions (K1,K2,K3, K7,K8, K10)

## 5  Experimental results

Experimental results are compared with estimates obtained from SHOCK [9]. A relatively large difference has been observed between the estimates and the experimental results notably regarding the values of peak overpressure. For instance, discrepancies as large as 70% have been observed. These major discrepancies have been imputed to the shape of the explosive charge. Indeed, the SHOCK software [9] considers a spherical charge whereas the trials were performed with cylindrical charges. In the configuration used for the trials, the reduced distance between the explosive charge and the gauges is rather small (less than 1 m/(TNT kg)$^{1/3}$) and the influence of the shape of an explosive charge is more important for the smaller the distances, see fig. 5 taken from references [8] and [11]. The comparison of the impulse values between trials and a priori estimates is about 27%, showing less influence of the charge shape itself.

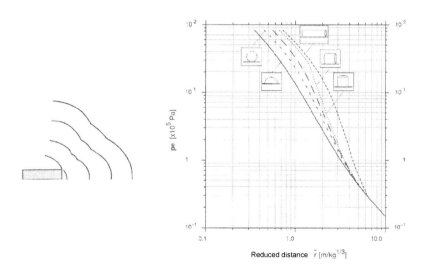

Figure 5:    Influence of the shape of the charge.

## 5.1 Position of the gauges

In this section, the different pressure profiles are presented considering the position of the gauge.

Figure 6 is the record of pressure vs. time history of the gauge K4, placed in the middle of the largest face of the experimental box. The maxima of pressure are obtained for this gauge, which is the closest gauge to the charge. It can be seen that the first overpressure peak is rather high, and followed by several peaks due to the different reflections.

Figure 7 is the record of pressure vs. time history of the gauge K9, placed in the middle of the smallest face of the experimental box. On the results, it has been observed that the largest overpressure peak appears a few microseconds after a first one, smaller, and is then followed by several peaks. This second

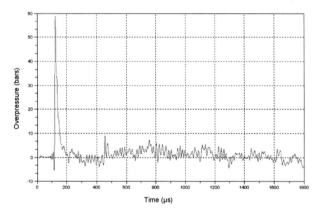

Figure 6:    Centre of the largest face (K4, 16 g eq TNT).

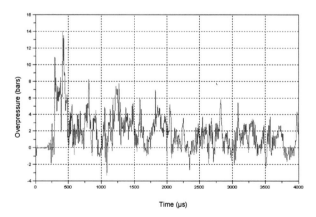

Figure 7:    Centre of the smallest face (K9, 16 g eq TNT).

overpressure peak is induced by the four simultaneous reflections on the walls surrounding the smallest face, which are at equal distance from the gauge.

Figure 8 is the record of pressure vs. time history of the gauge K5, placed in one corner of the room. It can be noticed that in this case, the first peak to appear is much less intense than the immediate second one. This particular gauge being near and equidistant from three surfaces of reflection, the first peak corresponds to the incident wave and the second to the combination of all the reflected waves on the adjacent faces. The following peaks are induced by the following reflections.

## 5.2 Scale factor

Considering the Crank-Hopkinson's scaling law, the pressure ratio for the two scale mock-ups should be equal to 1, and the impulse ratio equal to the scale factor, i.e. 0.63.

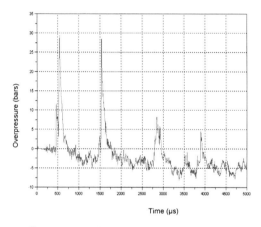

Time (μs)

Figure 8:     Corner (K5, 16 g eq TNT).

Table 1:     Ratio between peak overpressure at two different scales. The average was obtained considering all the gauges and all the trials.

| gauges \ masses | K1 | K2 | K3 | K4 | K5 | K6 | K7 | K8 | K9 | K10 | K11 | Average |
|---|---|---|---|---|---|---|---|---|---|---|---|---|
| 2g/8g | 0.934 | 1.555 | 0.882 | 0.721 | 1.370 | 0.900 | 0.445 | 0.911 | 1.018 | 1.035 | 0.763 | 0.958 |
| 4g/16g | 0.636 | 1.142 | 0.677 | 1.438 | 0.957 | 0.903 | 0.591 | 0.713 | 1.510 | 0.926 | 1.044 | 0.958 |
| 4g/16g | 1.201 | 1.386 | 1.173 | 0.684 | 0.581 | 1.354 | 1.330 | 0.857 | 1.176 | 0.865 | 0.739 | 1.186 |
| 4g/16g | 1.215 | 1.104 | 0.828 | 0.629 | 0.908 | 1.145 | 0.389 | 1.179 | 1.244 | 0.738 | 0.767 | 0.922 |
| 4g/16g | 0.870 | 0.810 | 0.867 | 1.192 | 0.894 | 1.115 | 0.715 | 0.938 | 0.980 | 0.615 | 0.608 | 0.873 |
| 4g/16g | 1.306 | 0.884 | 1.445 | 0.740 | 0.730 | 1.004 | x | 0.763 | 0.621 | x | 0.870 | 0.929 |
| 3g/12g | 0.878 | 1.904 | 1.049 | 0.377 | 1.122 | 0.747 | 0.475 | 0.798 | 1.010 | 0.826 | 0.796 | 0.908 |
| 3g/12g | 0.934 | 1.530 | 0.782 | 0.599 | 1.403 | 0.645 | 0.577 | 1.077 | 0.550 | 0.693 | 1.012 | 0.891 |
| 1g/4g | 1.041 | 1.291 | 0.573 | 0.729 | 1.245 | 1.311 | 0.891 | 0.864 | 0.977 | 0.837 | 1.719 | 1.043 |
| 2g/8g | 1.180 | 0.647 | 0.875 | 0.488 | x | 0.477 | x | 0.748 | 0.845 | 0.792 | 0.754 | 0.756 |
| 3g/12g | 1.268 | 1.097 | 1.000 | 1.271 | 1.133 | x | 0.641 | 1.429 | 1.086 | 0.838 | x | 1.085 |
| 4g/16g | 0.803 | 0.942 | 0.660 | 0.816 | 1.485 | 1.602 | 0.669 | 1.128 | 0.979 | 0.576 | x | 0.966 |
| | | | | | | | | | | Average | | 0.956 |

These results give good insurance that the Crank-Hopkinson's scaling law has been verified through the present internal blast loading experimental campaign.

The compared recordings of the profile pressures vs. reduced time, obtained for the explosion of 4 g (eq TNT) in the small model and 16 g (eq TNT) in the large model are drawn on figs 9 and 10. The good correspondence between the curves obtained at one scale and at another, confirms the good reproducibility of the tests even at different scales.

### 5.3 Gas pressure

For each size of mock-up and each mass of explosive, trials were performed with different type of openings in order to emphasize the influence of the gas pressure on the pressure profile measured on the mock-up faces. The recordings obtained for these three different configurations are plotted in fig. 11. During the first

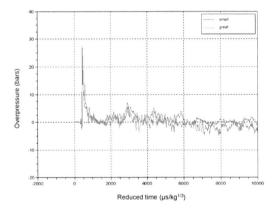

Figure 9:     Centre of greatest face (K4, 4 and 16 g (eq TNT)).

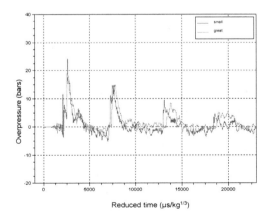

Figure 10:     Corner (K5, 4 and 16 g (eq TNT)).

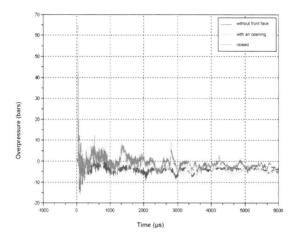

Figure 11:    Pressure for the three type of opening (4 g eq TNT).

milliseconds of the test, the pressure profile is the same in the three configurations; this part corresponds to the air blast propagation and reflections; after that, differences between the profiles begin to appear. The pressure profiles corresponding to the closed mock-up and the mock-up with one opening present more intense peaks than in the pressure profile observed for the open mock-up. Furthermore, the amplitude of these peaks is more important in the case of closed mock-up. This observation confirms qualitatively that the detonation gas pressure has an influence on the pressures measured by the gauges. Nevertheless, considering the effect of the drift of the gauge in the (0–200 bar) range, the evaluation of the gas pressure amplitude remains difficult to obtain out of the present experiments.

## 6    Conclusions

In order to conduct security studies for which the effects of blast waves in the vicinity of industrial sensitive installations need to be investigated, IRSN has developed an experimental set-up composed of a modular table, mock-up and pressure transducers. This experimental set-up is a support for non-destructive studies and dedicated to testing various shock wave propagations. The first campaign performed in the end of 2006 allowed to qualify the measurement chain and validate the concept of small-scale experiments. This validation was conducted through a free-field campaign. This campaign performed in 2009 allowed to validate the concept of small-scale experiments for an internal blast. In this framework, the results obtained allowed to check that the Crank-Hopkinson's "cube-root" law is verified. This campaign has also shown that improvements should be made in order to quantify the effects of gas pressure, notably requiring the use of some other sensor technology or range.

# References

[1]   Cheval, K., Loiseau, O. and Vala, V., Laboratory scale test for the assessment of solid explosive blast effects. *Structures under Shock and Impact X*, eds. N. Jones and C.A. Brebbia, pp. 63-72, 2008

[2]   Miura, A., Mizukaki, T., Shiraisji, T., Matsuo, A., Takayama, K. and Nojiri, I., Spread behaviour of explosion in closed space. *Journal of Loss Prevention in the Process Industries*, **17(1)**, pp. 81-86, 2004.

[3]   Keenan, W.A. and Tancreto, J.E., *Blast environment from fully and partially vented explosions in cubicles*, Civil Engineering Laboratory (Navy), Port Hueneme, CA, 1975.

[4]   Chan, P. and Klein, H.H., A study of blast effects inside an enclosure. *Journal of Fluids Engineering*, **116(3)**, pp. 450-455, 1994.

[5]   Zyskowski, A., Sochet, I., Mavrot, G., Bailly, P. and Renard, J., Study of the explosion process in a small scale experiment – Structural loading. *Journal of Loss Prevention in the Process Industries*, **17(4)**, pp. 291-299, 2004.

[6]   Sonoda, N., Hashimoto, A. and Matsuo, A., Influence of vessel geometry on the effect of explosion vent. *5th International Seminar on Fire and Explosion Hazards*, Edinburgh, UK, 2007

[7]   Kinney, G.F. and Graham, K.J., *Explosive Shocks in Air*, Springer Verlag: Berlin, 1985.

[8]   Baker, W.E., Cox, P.A., Westine, P.S., Kulesz, J.J. and Strehlow, R.A., *Explosion Hazards and Evaluations*, Elsevier: New York, 1983.

[9]   SHOCK, Naval Civil Engineering Laboratory, USA.

[10]  FRANG, Naval Civil Engineering Laboratory, USA.

[11]  Anet, B. and Binggli, E., *LS2000 Luftstoss phänomene infolge nuklearer und konventioneller explosionen.*

# Laboratory simulation of blast loading on building and bridge structures

M. M. Gram[1], A. J. Clark[1], G. A. Hegemier[2] & F. Seible[2]
[1]MTS Systems Corporation, Eden Prairie, MN, USA
[2]Department of Structural Engineering, Jacobs School of Engineering,
University of California, USA

## Abstract

There is a need to modify structural elements of buildings and bridges in order to improve the response to blast loads. Testing the modifications has been a major stumbling block to advances. Failure modes resulting from a 1 to 10 ms blast impulse are substantially different from the failure modes obtained with conventional laboratory structural tests. Field tests using actual explosions are expensive, are not repeatable, and the fireball makes visual and video viewing impossible and reliable real-time data very difficult to obtain. The University of California, San Diego and MTS Systems Corporation were funded to design and build a system able to perform laboratory blast simulations. The system, which was completed in 2005 at UCSD, has been used to test modified and unmodified reinforced structures including: concrete columns 355 mm square by 3 m high; and concrete block walls 200 mm thick by 3 m high. By January 2006, over 20 specimens were tested, with specimen failure modes matching the failure modes from field tests. The system uses impact loading to produce a 2 ms pulse with a typical peak pressure loading of 35 MPa and an impulse of 14 kPa-s over the surface of the column. The system allows observation of the test in person and by high-speed video, as well as successful instrumentation of the specimen. This paper describes the blast simulation system and its capabilities.
*Keywords: blast simulator, blast hardening, blast mitigation, infrastructure protection.*

# 1 Background

Terrorist attacks world-wide have demonstrated our vulnerability to threats in the form of conventional and improvised explosives. Such threats have mandated the development, validation and deployment of blast resistant new construction, and hardening retrofit techniques for existing structures. Efforts to accomplish objectives in this area are underway at the Powell Structural Research Laboratories at the University of California, San Diego (UCSD) as part of the international TSWG (Technical Support Working Group) blast mitigation program.

The development and validation of blast resistant design methodologies for new and existing structures requires high fidelity data that is repeatable and available in sufficient quantities. A similar statement applies to the problem of validating and/or improving design or blast physics computational tools.

It is evident that field testing of structural components and assemblages are necessary to complete this process. As a result, UCSD has been performing a range of field tests at sites such as the Energetic Materials Research and Testing Center (EMRTC) of New Mexico Tech (our partner in the TSWG blast mitigation program) and the Naval Air Warfare Center Weapons Division at China Lake, California.

It is also evident, however, that field experiments are limited by constraints on costs and time. In addition and perhaps more important, reliable real-time field data are very difficult to obtain and the fireball renders visual and high speed video viewing of failure processes virtually impossible.

In an effort to expand the quality and quantity of blast experiments in an affordable and timely manner, UCSD and MTS were tasked and funded by the TSWG to design and build a system able to perform blast load simulations on typical structural components in a laboratory environment. The result is the UCSD Blast Simulator, the world's first hydraulically driven device that simulates explosive loads without the use of explosive materials, and without a fireball. In what follows, the design and performance of the Blast Simulator are described.

# 2 Concept

In field-testing, an explosion produces a pressure pulse that can be measured using pressure transducers distributed on the test specimen. The impulse (momentum) is calculated from the pressure data. The challenge was to develop a test system to reproduce that impulse by a method that was suitable for a laboratory. In response to that challenge, UCSD and MTS, worked together, to develop the Blast Simulator. The technology used in the Blast Simulator was developed by MTS Systems for production of test equipment requiring high energy impacts: artillery firing simulators, automotive crash simulators, shock testing equipment, and high rate material test systems.

The concept of the Blast Simulator is to produce an impulse by impacting the specimen with a mass. An elastomeric spring between the impacting mass and the specimen distributes the force over the uneven surface of the specimen. The impacting mass is brought up to velocity. The impact transfers the energy and momentum from the impacting mass to the specimen. The spring rate of the elastomer determines the duration and the peak force of the impact pulse.

There is a need to study and test many types of structures, so the Blast Simulator must be adaptable. By using multiple actuators with an impacting mass attached to each, the system can be configured to test a wide variety of structures. The requirement for controlling the momentum delivery system then must include precise impact timing, as well as, precise impact velocity, in order to simulate an actual blast.

## 3  The blast impulse

A standard technique for analyzing structures subject to a blast is to use the impulse (time integral of force); this technique is considered accurate for structures having a response time greater than 3 times the length of the impulse (response time is defined as the time to peak displacement) [1]. Therefore, the Blast Simulator elastomeric springs must provide a short pulse similar in length to those measured in field tests.

Table 1:      Technical specifications for the Blast Generator.

| Blast Generator | |
|---|---|
| Quantity | 4 |
| Maximum energy (each) With mass:   50 kg (110 lb) 100 kg (220 lb) 200 kg (441 lb) 400 kg (882 lb) | 30 kJ 51 kJ 76 kJ 101 kJ |
| Maximum velocity With mass:    50 kg (110 lb) | 34 m/s (1300 in/s) |
| Repeatability of velocity | 4% or 0.1m/s whichever is greater |
| Simultaneous impact | Within 0.002 s |
| **Impacting Mass** | |
| Mass range (bolted to BG rod) | 50 to 400 kg (110 to 882 lb) (includes mass of piston) |
| Mass range (free mass) | 10 to 50 kg (22 to 110 lb) |
| **Impact Spring** | |
| Impulse time | 0.5 to 5 ms |
| Linearity (non-linearity) | Spring rate to increase with increasing deflection to prevent an overly abrupt impulse initiation. |
| Mounting of impact spring | Bolts to face of impacting mass |

A strategy for creating a simulation equivalent to an actual blast when either pulse is too long is to use finite element analysis to calculate a correction factor. A finite element analysis of the blast pulse on the structure determines the work put into the structure by the blast, then an impact velocity can be determined (using finite element) that subjects the structure to the same work.

In the simulation, the mass attached to the Blast Generator is accelerated to a velocity. The resulting momentum equals mass × velocity. Momentum from the impactor is transferred to the specimen during the impact. The momentum transferred is a function of the mass of the impactor, the mass of the specimen, and the damping of the elastomeric spring. The actual momentum or impulse delivered to the specimen during a test is calculated from accelerometer data and Blast Generator pressure transducer data. The specifications for the Blast Generator are listed in Table 1.

## 4  Blast Generator

In order to develop the energy or momentum to test the specimen, the Blast Generator actuator must accelerate the impacting mass to a velocity up to 30 m/s (1200 in/s) within a stroke of about 1 m (40 in). This hydraulic actuator, together with the control valves, accumulators and transducers, is referred to as a Blast Generator (BG). A schematic diagram of the BG is shown in fig. 1.

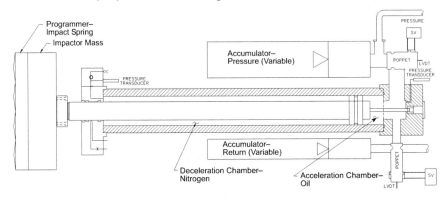

Figure 1:    Blast Generator schematic.

The actuator can produce a maximum force of about 250 kN (56000 lbf) to accelerate the impacting mass. To provide the peak flow for the test, nitrogen is compressed in the pressure accumulator to store a volume of high-pressure oil. The servo-controlled high-flow valve controls the oil flow into the acceleration port of the actuator. Potential energy in the accumulator is converted, in the actuator, to kinetic energy in the impacting mass. A smaller servo-controlled valve controls flow out of the acceleration port to retract the actuator after the impact. The force to retract the actuator is supplied by pressurized nitrogen gas that fills the retract, or deceleration, chamber of the actuator. In a conventional actuator, oil is used in the retract chamber; in the BG, nitrogen gas is used to minimize energy loss.

The time to accelerate the mass to a stable impact velocity is typically 40 ms. A close-coupled, pressure-line accumulator incorporated into the BG assembly provides the fast flow response required to supply the needed energy. A close-coupled, return-line accumulator accommodates the flow from the retract valve.

The two high-flow valves are servo controlled, using the LVDT feedback on valve position. During the test, the pressure valve opening is controlled precisely to form the control orifice needed to achieve the desired impact velocity and impact timing. During setup the two valves are operated together providing position control for the actuator.

Pressure transducers provide readings of all the actuator pressures to allow precise setup and to measure actuator forces during impact. Each actuator position is measured and controlled using externally mounted Temposonic® magnetostrictive position transducers. The differentiated stroke transducer output provides an accurate measure of impact velocity.

A steel plate, bolted to the end of the piston rod provides the impacting mass. An elastomer spring is mounted to the mass. The impacting face of the elastomer spring is contoured to provide an initially low contact spring rate that progressively increases; this contouring reduces excitation of high frequency vibrations in the impacting mass and the specimen. The elastomer spring is shown in fig. 2.

Figure 2:     Elastomer spring.

## 5   Control system

The computer based control system controls BG positions during test setup. During the test, it provides the valve commands to generate the desired impact velocity and timing on the four BGs. After the test, it retracts the BGs and returns them to a safe state. A computer model of the BG is used, for each test, to determine the setup values for oil and nitrogen pressures, initial position, and valve command profile to achieve the desired impact velocity.

The control system was initially designed to run tests with all actuators impacting at the same velocity. After conducting tests for several months, the

UCSD researchers saw a need for tests simulating a bomb close to the column specimen with higher impulse (velocity) at the lower end of the specimen and reduced impulse on the higher BGs. MTS was able to accommodate this need with a minor change to the software. The system is able to run tests with impact velocities varying over a 3:1 range, still maintaining simultaneous impact.

## 6  Instrumentation and data acquisition

Control and setup transducers include three pressure transducers on each Blast Generator, LVDT stroke transducers on the control valves, and a Temposonic stroke transducer used to control actuator position and measure velocity at impact. Accelerometer(s), mounted on the impacting mass, record impact data for calculating force and impulse. Output of all the BG control system transducers is digitally recorded at 4 kHz for each test. The accelerometer output is recorded at 800 kHz.

The specimen instrumentation system was configured by UCSD. Transducer selection varies to suit the test requirements and includes accelerometers, stroke transducers, and strain gages. Data acquisition, triggered by the control system, can record 52 channels of 14-bit data at 1 MHz. The high-speed data acquisition can also record data from any of the control transducers, when desired.

Two Phantom version 7.1 high-speed video cameras, capturing 5000 frames per second, provide a graphical capture of specimen failure, and can also provide velocity and deflection measurements of the specimen.

## 7  Measuring the impulse

The force on the elastomeric spring acts equally on the specimen and the impacting mass. Forces acting on the impacting mass are: specimen force on the elastomer pad, friction on the bearings, and BG internal forces. Thus the specimen force is the inertial force (acceleration times impacting mass), minus friction, plus BG force. The impulse measurement is as follows:

- Accelerometers mounted on the impacting mass measure the acceleration during the impact
- Friction force can be calculated or measured (approximately) and subtracted. It is very small and is usually ignored.
- The BG is controlled to provide low acceleration at the point of impact. This holds the BG forces relatively low during impact–generally <5% of inertial force.
- The pressure transducers measure the internal pressures in the BG actuator, and the force is calculated. The speed of sound in steel and oil causes a delay of about 1 ms between the transducer and accelerometer data. This delay is calculated (based on actuator position at impact) and the force during impact is recorded and added as a correction to the inertial force.
- The resulting force, on each impactor, is integrated over the time of the pulse to find the impulse delivered to the specimen.

Accelerometers mounted at various locations on the specimen can be used to calculate specimen velocity and displacement during the test.

## 8  Foundation

The isolated foundation has a reinforced concrete reaction wall at each end, as shown in fig. 3. The fixed reaction wall has a steel mounting plate for mounting the BGs. For tests to date, the BGs have been mounted in a vertical column utilizing t-slots in the wall to provide adjustment. The mounting plate allows other BG mounting locations as required.

The moveable reaction wall provides a moveable link system for the upper end of the specimen. The moveable reaction wall is re-configurable. It is constructed from large reinforced concrete blocks that are stacked and post-tensioned to the foundation. Fixtures have been built to allow simulations of end conditions on the specimen matching those in actual structures.

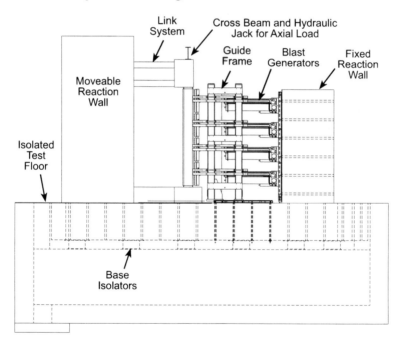

Figure 3:    Foundation.

Guide rails are used to support the weight of the impacting masses and keep them aligned with the specimen. The guide rails are attached to an adjustable frame that allows spacing between BGs to be adjusted. It also adjusts to accommodate several sizes of impacting masses. The stroke transducers are mounted to the guide frame, and provide a direct reading of the position of the impacting mass. A photo of the guide frame can be seen in fig. 4.

Figure 4:     Guide frame (shown with broken specimen).

## 9  Infrastructure

The blast simulator is part of the Englekirk Center, a multi-hazard research facility of the University of California, San Diego about eight miles east of the main UCSD campus.  In addition to the Blast Simulator, the center includes two other major research projects: The world's largest outdoor earthquake simulation table–the LHP Outdoor Shake Table; and the Soil Foundation Structure Interaction (SFSI) program, which includes a large laminar soil shear box and two refillable soil pits.

The three facilities share a hydraulic power supply that includes an accumulator bank for providing peak hydraulic flow of 40,000 l/m (10,000 gpm).

The very temperate San Diego climate allows test equipment to be operated outside throughout the year.  By locating the equipment outside, full scale specimens can be built and installed using conventional construction techniques.  The site has a 55-ton truck-crane available for all the test rigs.

The Blast Simulator is connected to the hydraulic power supply and uses 34 MPa (5000 psi) pressure for primary flow, as well as 21 MPa (3000 psi) for pilot pressure.  The accumulator bank flow is also available to simulate after-the-blast loading on the specimen.

The Blast Simulator foundation is sited so that the BGs are located inside the building, and the specimen is outside the building. An outdoor preparation area adjacent to the system is used for building and curing specimens. As one test is completed, the truck crane can quickly move a new specimen into position for mounting, post tensioning, instrumenting, and testing.

## 10  Blast Simulator performance

Performance of the Blast Simulator has been evaluated to-date using BG-on-BG impacts, and column and wall test specimens loaded with groups of three or four BGs at impact velocities from 1.5 m/s (4.9 ft/s) to 30 m/s (98 ft/s). Below, we present a subset of the column test data as an indicator of performance.

### 10.1 Description of test specimen and test setup

The columns to be discussed are reinforced concrete (RC) with full-scale dimensions of 3.28 m (129 in) in length and 355.6 mm (14 in) square. These are loaded with four BGs distributed over the column length as depicted in fig. 2. The impact mass plates measure 355.6 mm (14 in) wide by 762 mm (30 in) tall and have a mass of 243 kg (535.7 lbs). The reaction fixtures are similar to those used in previous quasi-static tests of blast resistant column designs conducted by UCSD [2, 3]. At the column base, the footing is post-tensioned to the test floor in an effort to restrain all degrees of freedom. At the top of the column, a link system allows the column to move vertically while providing lateral and rotational fixity. Vertical axial load is applied at the top of the load stub by three hydraulic jacks with a mechanical lock-off. The jacks react against a steel frame. This test setup is intended to simulate the boundary and loading conditions associated with an actual building configuration. (Details of the test setup can be found in Seible and Hegemier [4] and Hegemier et al. [5].)

### 10.2 Energy deposition signature

The force-time histories associated with the mass plates plus programmer-specimen impacts, as determined by acceleration-time histories, reflect a pulse width of 2 to 4 ms (0.002 to 0.004 s) depending on the programmer type and the impact velocity. In addition, simultaneity of impacts (see fig. 5) can be achieved in a range of 0.3 ms to 2 ms depending on the test settings and conditions.

### 10.3 Impulse histories

Typical impulse histories are shown in fig. 6 along with the overall (averaged) impulse. In general, the overall impulse can be delivered in 3 to 5 ms with amplitudes up to 20.7 kPa-s (3 psi-s). An impulse of 14 kPa-s (2 psi-s) is representative of a "car bomb at curb side".

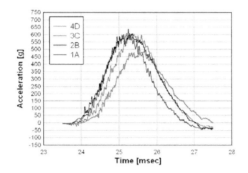

Figure 5:    Typical simultaneity of BG hits (0.3 ms spread).

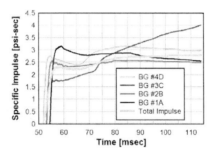

Figure 6:    Impulse histories.

Figure 7:    From left to right: RC column (field test), RC column (blast simulator test), CFRP wrapped column (field test), CFRP wrapped RC column (blast simulator test).

## 10.4 Comparison of laboratory and field tests

Comparison of post-test laboratory and field data from similar tests conducted on similar test specimens have revealed excellent correlation of impulse, deformation, and failure mode, thus showing the blast simulator accurately simulates live explosive loads. Example comparisons of laboratory and field impulse and failure modes are shown in figs. 7 and 8, respectively, for a "car

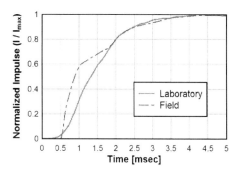

Figure 8:      Impulse comparison.

Figure 9:      Damage evolution of an as-built column under blast simulator impact.

bomb at the curb" scenario. Finally, it is emphasized that the absence of the fireball associated with the field event allows one to follow the specimen failure-time history. This is dramatically illustrated in fig. 9 which is a frame sequence from high speed video record of a RC column subjected to a 15.8 kPa-s (2.3 psi-s) event. In the field, such information is typically obscured by the fireball and the debris field.

## References

[1]     Joint Departments of the Army, the Navy, and the Air Force, *Structures to Resist the Effects of Accidental Explosions*, TM 5-1300/NAVFAC P-397/AFR 88-22.

[2]     Hegemier, G., Seible, F., Rodriguez-Nikl, T., Lee, C., Budek, A.M., & Dieckmann, L., *FRP-Based Retrofit Strategies – Laboratory Tests on Rectangular RC Columns*, Report No. SSRP-2002/04, UCSD, Department of Structural Engineering, 2002.

[3]     Hegemier, G., Seible, F., Lee, C. & Rodriguez-Nikl, T., *FRP-Based Retrofit Strategies – Laboratory Tests on Rectangular RC Columns – Part II*, Report No. SSRP-2002/17, UCSD, Department of Structural Engineering, 2003.

[4]     Seible, F. & Hegemier, G., *Shakedown and Commissioning Final Test Report*, TSWG Contract Deliverable, 2005.

[5]     Hegemier, G., Seible, F., Rodriguez-Nikl, T. & Arnett, K., Blast mitigation of critical infrastructure components and systems. *Proceedings of the Second fib Conference*, Naples, Italy, 2006.

# Non-ideal explosive performance in a building structure

K. Kim[1], W. Wilson[1], J. Colon[1], T. Kreitinger[1], C. Needham[2],
R. Miller[2], J. Orphal[2], J. Rocco[2], J. Thomsen[2]
& L. V. Benningfield[2]
[1]*Defense Threat Reduction Agency, USA*
[2]*Applied Research Associates, Inc., USA*

## Abstract

The performance of a non-ideal explosive has been investigated in a reinforced concrete 2-room test structure with two different window and door configurations: (1) without and (2) with doors and a window. Pressure records at different locations of the two-room test structure for the two different configurations have been compared with each other and 3D CFD numerical simulations. Reaction efficiencies of fuel gases and aluminum particles in the explosive detonation products, especially the aerobic reaction with air, as functions of time have been estimated for the two cases and then compared with each other.

*Keywords: non-ideal explosives, aluminum particles, aerobic reaction, anaerobic reaction, tests, CFD simulations, payload-structure interaction.*

## 1 Introduction

Generic non-ideal explosives are fuel-rich in detonation reactions. Much of the fuel, in the form of aluminum particles and some detonation products, will react with other detonation products during an anaerobic reaction phase in a fireball. However, typically, there is some leftover fuel that has a potential to react further with the surrounding air (aerobic reaction) at a later time. If it does, in what timescale does it react? Does it add to the strength of blast waves inside a structure? Would it work if the structure is open to the outside (doors and windows are removed from their openings)? Inside the structure, how much blast enhancement can be obtained if the structure is marginally and temporarily contained by responding doors and windows? If so, how much and through what mechanisms are these enhancements realized? These are the questions to be addressed by this paper.

Two identical tests, with the exception of the presence or absence of responding doors and windows, were conducted and then numerical simulations were performed to aid in determining the reasons for differences between the two cases, if any.

## 2 Test description and results

Test Charges: Each test charge was cylindrical having a diameter of 11.0 cm, a length of 24.1 cm, and a nominal mass of 3.73 kg. The mild steel cases had a thickness of 0.625 cm on both the cylindrical portion and the end caps. The end caps were bolted to the cylindrical portion of the case resulting in a nominal total case mass of 5.45 kg. The charges were end detonated with a detonator and a 50 g cylindrical high explosive booster (3.175 cm diameter × 3.62 cm length) embedded in the main charge.

### 2.1 Test-bed configuration

The reinforced concrete test-bed configuration is shown in fig. 1. This 2-room configuration is one half of a 4-room test structure with the other half of the structure being separated from the two rooms used in these tests using steel non-responding doors (sealed hallways shown in black in fig. 1). The walls are 0.91 m thick steel reinforced concrete and do not deform during tests. The rooms have nominal 3.66 × 4.57 × 2.44 m inside dimensions. The windows are nominally 1.07 m wide by 0.91 m tall, and doors are nominally 1.07 m wide by 2.13 m tall. The charges were placed parallel to floor and parallel to the interior wall as indicated in fig. 1. The center-of-mass of each charge was located 1.09 m from the interior wall, 0.91 m from the floor, and 0.91 m from the wall opposite the window. The charges were end detonated from the end facing the window wall (bottom end in fig. 1). The source room interior was clad with 1.27 cm thick steel with an additional 1.27 cm steel cladding placed on the floor, walls, and ceiling in the fragment pattern of the test devices.

There were many gages employed inside and outside of the structure (see [1] for further details), but in this paper they are not shown. One of the static pressure gages is located in the middle of the source room in the ceiling (red triangle in fig. 1) and there are five other wall-mounted gages in the same room (blue squares in fig. 1, and two of them are located at position A at different heights). The gage selected for illustration in this paper is indicated as position "A" in the diagram and is located 76 cm from the floor in the corner of the source room opposite the corner where the charge was placed. This gage location is typical for pressure measurements inside the source room and has been documented to produce impulse readings within ±3% for a number of shots with identical charges. In one test (Test 1), all doors and windows were removed from the structure (leaving the structure open to the outside), and in the other

(Test 2), two plywood doors and a plastic-film-covered window were placed as shown in fig. 1.

For Test 2, a 1.3 cm thick plywood door, slightly smaller than the steel opening (by ~0.3 cm), was mounted in the exterior doorway flush with the interior wall using three hinges on the left (as viewed from the interior). The right side of the door was secured with screws to three 5.1 × 10.2 × 7.6 cm wood blocks epoxied to the inside of the door frame. A 1.3 cm plywood door between the source and adjacent rooms was flush with the inside of the adjacent room and was mounted with hinges and wood blocks as for the source-room door (hinges on the left as viewed from the adjacent room).

For Test 2, the window opening was fitted with standard 0.64 cm thick glass with a window film on the surface facing out of the structure. The surface of the glass was set just outside the inner surface of the room, mounted in a 5.1 × 10.2 cm wooden frame. The wood frame was supported from behind by 2.54 cm angle steel tack-welded to the window opening. After the test, the steel was still in place but the wood frame had been ejected.

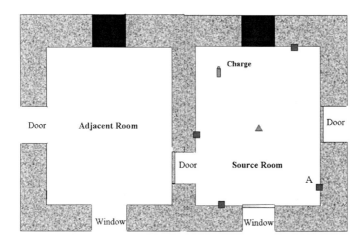

Figure 1: Test bed configuration and gage locations – shown with doors and window in place (Test 2 configuration).

## 2.2 Test results

Figure 2 shows the comparisons of typical pressure records taken for the two test configurations. The solid blue lines show the static pressure for Test 1 with unrestricted openings, and the broken red lines show the same for Test 2 where the doors and window were in place. From the video results, it was concluded that the plywood doors were ejected as intact units within a few ms after the impact of the first blast wave. The behaviour of the window was not clear, but the mass of the glass probably provided a temporary confinement effect within the structure. It is this temporary containment effect by the doors and window that influence the explosive performance in the structure.

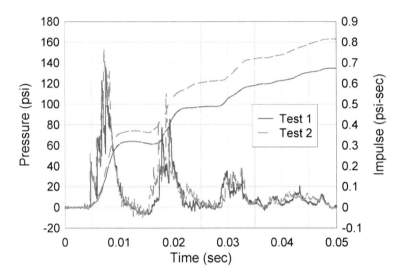

Figure 2:    Pressure and impulse records for Test 1 (solid blue) and Test 2
             (dashed red).

Qualitatively, the two tests exhibited similar behaviours, showing similar peaks and valleys. There were several shock wave vibrations in the room due to geometric effects within the first ten to twenty ms. The first pressure peaks were almost the same in magnitude for both cases, as they should be since the first peak travels directly from the detonation of the charge to the sensor without being influenced by the structure surfaces. Subsequent peaks are reflected waves and some reflected secondary waves can be larger than the first waves.

A major difference between the two tests is that in the case where doors and the window are in place (Test 2), the pressure does not decay as fast as the case where they are absent (Test 1). This can be clearly seen in the impulse curve comparisons. In the source room wall locations, the difference is about 15–20% in impulse. In the adjacent room, although not shown here, most of the gages show a difference of about 10–20% in impulse.

The difference of 15–20% in impulses may not seem significant, especially if one examines the peak pressure only. A non-ideal explosive is not expected to show as much difference since the majority of aluminum fuel particles are expected to react in anaerobic reactions within the detonation product gases, and only a small portion of aluminum particles are expected to react with the surrounding air. Although the difference is small, it is well outside the experimental error and is useful in understanding the phenomenology involved. Also, this difference is directly related to the kinetic energy of secondary debris in conventional deformable structures and can play a very important role.

The main question is: Is this difference coming from a more localized distribution (because of the slight confinement effect of the doors and the window) of the same total energy from the detonation of the charge? Or, does the temporary confinement effect of the doors and the window affect the reaction efficiency of the charge as well? To be more precise, does the temporary confinement enhance the aerobic and anaerobic reaction efficiencies of a non-ideal explosive and therefore its total energy output? How long does the confinement effect last? From the pressure data, it might be conjectured that there must be some residual reaction going on within the structure. In order to gain some insight into answering these questions, numerical calculations were made simulating the two test cases.

# 3 Numerical simulations with SHAMRC

## 3.1 The CFD model

The SHAMRC[2] CFD non-ideal explosives model attempts to incorporate first-principles physics and chemistry models into a CFD code. The goal of this model is to predict post-detonation environments for non-ideal explosives given the target geometry, weapon specifications, and explosive properties of the formulation. The model allows the detonation products and embedded aluminum particles to evolve over time, with the particles being heated and cooled by the surrounding gasses and accelerated by the drag of the blast flow gases within the target geometry. If the temperature of the aluminum particles reaches 2050 K, and if there is oxygen present nearby, the aluminum particles are allowed to ignite and release energy. The reactions considered by the model are a "first-order" approach to capturing the post-detonation energy release of non-ideal explosives.

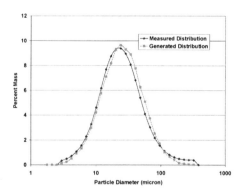

Figure 3:     Aluminum particle size distribution.

In the model, aluminum particles, perhaps the most important fuel present in many non-ideal explosives, are treated as being spherical with the same surface-to-volume ratio as the actual particles. The model approximates the size distribution used in a particular explosive mixture. For example, fig. 3 shows a comparison between the MDX-81 particle size distribution used in some aluminized explosives and the distribution generated by the model. The size distribution of aluminum particles is important because the heating, and therefore the ignition and burning of the aluminum particles are strongly dependent on the particle size.

The amount of carbon, water, methane, and "other burned explosive materials" released during the passage of the detonation front are estimated using the CHEETAH program provided by Lawrence Livermore National Laboratory. After the detonation products are released, the model allows them to react with atmospheric oxygen and aluminum particles. The following reactions are considered:

$$2Al + 3H_2O \rightarrow Al_2O_3 + 3H_2$$
$$4Al + 3O_2 \rightarrow 2Al_2O_3$$
$$C + O_2 \rightarrow CO_2$$
$$2H_2 + O_2 \rightarrow 2H_2O$$
$$CH_4 + 2O_2 \rightarrow 2H_2O + CO_2$$

These equations represent the estimated primary reactions that occur after detonation and account for most of the post-detonation energy released based on this model. $CO_2$ in the detonation products is not allowed to react while water vapour is. It should be noted CHEETAH predicts that hydrogen gas is not found in abundance after the detonation of a typical organic explosive, thus hydrogen gas is only available for combustion after aluminum has reacted with water. The model allows the hydrogen, released by the reaction of water with aluminum, to react with atmospheric oxygen.

In the SHAMRC calculations, the doors and window were represented as high density fluids with their density equal to that of the doors and windows. No material strength was given to small latches, screws, mounts, or door and window materials in the calculation. This approximation has been used for "responding" objects in past calculations with relatively accurate results. The doors and window of the structure may have been propelled out of the structure as single units, at least for the first few ms. However, it was deemed that any error from this "dense fluid" approximation would not alter any conclusions of the paper qualitatively.

Figures 4 and 5 show the comparison between measured and calculated pressures at gage location A. There are some differences between the calculations and the measurements. The calculated arrival time of the first peak is slow by 0.6 ms (in the figures, the calculated values are shifted by 0.6 ms) and the calculated second major peak is also slower. The error may come from inaccurate description of initial detonation. Subsequent smaller peaks are somewhat out of phase and these differences may be due to the imprecise treatment of doors and windows. Fig. 4 (Test 1) shows better comparison between measured and calculated pressures than fig. 5 (Test 2). It is conjectured that, in Test 2, hinges, mounts and such that were ignored in the calculation may play some role in holding the confinement slightly higher, and therefore the measured values are somewhat higher than the calculated.

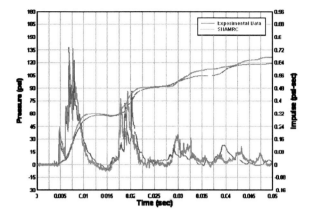

Figure 4:    Measured  and  calculated  pressures  for  Test  1  (unrestricted openings).

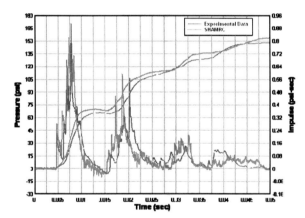

Figure 5:    Measured and calculated pressures for Test 2 (doors and window in place).

However, the overall comparison of figs. 4 and 5 is very good and this type of good  correlation  has  been  demonstrated  for  other  non-ideal  explosives  and thermobarics  in  a  number  of  other  tests.    It  is  therefore  assumed  that  the SHAMRC model is emulating most of the physics and chemistry fairly well even though it uses a fairly small set of chemical reactions and uses the "heavy fluid" approximation for responding objects.

## 3.2  Combustion efficiency of aluminum particles

One  possibility  for  the  increase  in  pressure  due  to  the  extra  confinement provided  by  the  responding  doors  and  window  is  aluminum  combustion efficiency increase for the non-ideal explosive.  Other reactions are assumed to occur as soon as mixing (with air) takes place whereas the aluminum reaction additionally requires that the particle temperature reaches its ignition temperature. Since the confinement increases the temperature inside the structure, aluminum reaction will benefit from the confinement, but not others. Figures 6 and 7 illustrate the combustion efficiency of aluminum predicted by the model for the two test conditions including the contributions of aerobic and anaerobic combustion of aluminum.

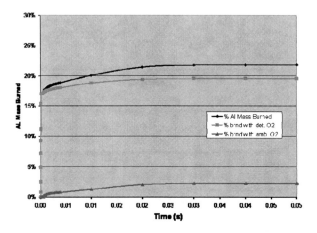

Figure 6: Aluminum combustion for Test 1 (unrestricted openings).

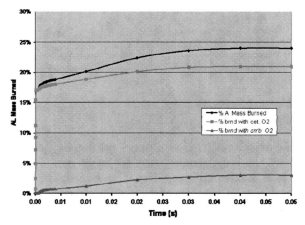

Figure 7: Measured and calculated pressures for Test 2 (doors and window in place).

An interesting observation drawn from figs. 6 and 7 is that much of the aluminum does not burn. The total amount of aluminum reacted is only about 22–24% (black lines). Of this, about 17% reacts very quickly (the nearly vertical line near time zero) after the detonation in an anaerobic reaction. The anaerobic reaction afterwards accounts only for 2 or 4% respectively and the aerobic reaction accounts only for 2 or 3% respectively, depending on the two test configurations.

The total aluminum reaction is increased only by 2% by the marginal confinement for this explosive. Enhancement of total energy release (not shown in the paper), is likewise increased by about 2% only. A large amount of post detonation reaction actually comes from carbon reaction (not shown) for this explosive. Most of the post detonation reaction for carbon is completed by about 30 ms, indicating that mixing with air takes that amount of time.

The aerobic reaction enhancement (the lower blue lines in figs. 6 and 7) is slightly less than that of the anaerobic reaction (the middle red lines). The fact that only a small amount of aluminum reacts at all indicates that aluminum does not get mixed with air in a short enough time when the temperature is high. Late time anaerobic reaction as well as aerobic reaction of aluminum takes about 30 ms (figs. 6 and 7). After 30 ms, the temperature of the air is no longer high enough to ignite the aluminum particles. A large amount of available energy stored in the explosive, therefore, is wasted during these events. Even some of the energy released by other fuel such as carbon may not contribute significantly to the enhancement of useful impulse, if they are released at later times when pressure peaks gets smaller.

Although not shown in the paper, the model predicts the door to the exterior door to begin moving at about 8 ms and is well outside the structure in about 20 ms, which correlates fairly well with the video data.

## 4 Conclusions

It has been demonstrated that the performance of non-ideal explosives (pressure, impulse and reaction efficiency) depends on target configurations. CFD calculations that closely approximate the measured pressure profiles indicate that the aluminum reaction efficiency is slightly enhanced by a temporary and marginal confinement such as responding doors and windows. This model also indicates that much of the aluminum in the explosive does not burn in these configurations. One reason may be attributed to limited mixing of aluminum particles with the ambient air at early times. Another reason may be the very high ignition temperature (2050 K) of the aluminum particles.

## References

[1] Colón, J.E., Sofran, G., Wilson, W., Babcock, S., Benningfield, L.V. and Kaneshige, M., *Enhanced Blast Standardized Test Program*, SAVIAC 75th Symposium Proceedings, Virginia Beach, VA, 2004.
[2] Watry, C., Needham, C., Perry, J. and Schneider, J., *Thermobaric ACTD Phenomenology and Explosive Model Development Report – Phase II*, Applied Research Associates, Albuquerque, New Mexico, 2004.

# Aerodynamic damping and fluid-structure interaction of blast loaded flexible structures

M. Teich & N. Gebbeken
*University of the German Armed Forces, Munich Campus, Germany*

## Abstract

This paper analyses the effects of air-structure interaction and derives a new coupling model for systems subjected to weak blast loads. While the coupling effects are negligible for typical steel or concrete structures, they may dominate the dynamic response of lighter and more flexible systems like membranes, blast curtains or cable facades. For these systems, a classical decoupled analysis, i.e., neglecting the influence of the surrounding air, might significantly overestimate the deflections and strains. The results of parameter studies are presented, and recommendations for the blast resistant design of flexible protective structures are given.
*Keywords: aerodynamic damping, blast-structure interaction, flexible protective structure.*

## 1 Introduction

The military has extensive knowledge and experience in protective design against blast and weapon effects. With increasing terrorist threats this knowledge is transferred to civilian applications [1]. Military protective structures like bunkers are often build with highly reinforced concrete. At the same time, civilian architectural trends head towards transparent, light and flexible structures like large cable net facades or membranes. Military methods and procedures, e.g. [2, 3], which are valid for the blast analysis of heavy and massive structures cannot be used in these cases. This paper contributes to a better understanding of the interaction of blast waves with light and compliant systems.

In the following Section 2, aerodynamic damping effects are illustrated. Then, in Section 3, we derive an analytical single degree of freedom (SDOF) fluid-structure interaction (FSI) model valid for the analysis of weak blast waves with linear or nonlinear structures. In Section 4, some parameter studies show the influence of aerodynamic damping and FSI on deflections and reflected pressure waves. Section 5 summarizes the main points of this paper and gives an outlook into further areas of research.

## 2 Aerodynamic damping

Aerodynamic damping is the result of energy dissipation of the surrounding air. While this effect is very small and usually neglected in standard protective design, it is taken into account for the design of structures subjected to strong wind loads and for flow-induced vibration analysis [4].

The basic principle of aerodynamic damping is illustrated in Figure 1. The plate with the specific mass $m$ and the velocity $\dot{x}$ impacts the air behind the system. This results in the propagation of an air wave moving with sound speed $c$ to the right. In the time span $\Delta t$ the wave travels the distance $\Delta x = c\Delta t$. The condition of continuity requires that the particle velocity $u_p$ is equal to $\dot{x}$. Thus, the affected air has the specific mass $m_{\mathrm{air}} = \rho\Delta x = \rho c\Delta t$ and the momentum $\Delta i = m_{\mathrm{air}}u_p = m_{\mathrm{air}}\dot{x} = \rho c\Delta t\dot{x}$. For $\Delta t \to 0$, the air pressure corresponds to $\Delta i/\Delta t = \rho c\dot{x}$. However, this is a linear approximation assuming that the particle velocity $u_p$ is constant in the time span $\Delta t$ and neglecting complex reflection phenomena and pressure relief effects at building edges and corners [5]. To take account of these effects, an experimentally determined drag coefficient $c_D$ is included in the determination of the aerodynamic damping pressure,

$$d_a = c_D \rho c \dot{x} \tag{1}$$

which can also be written as

$$d_a = 2m\omega\zeta_a\dot{x} \tag{2}$$

for linear systems following classical structural damping notation. In this case, $\zeta_a$ denotes the aerodynamic damping ratio depending on the air properties density

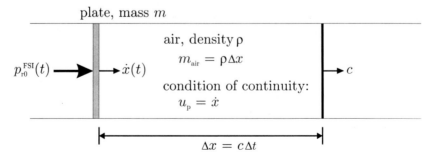

Figure 1: Basic principle of aerodynamic damping.

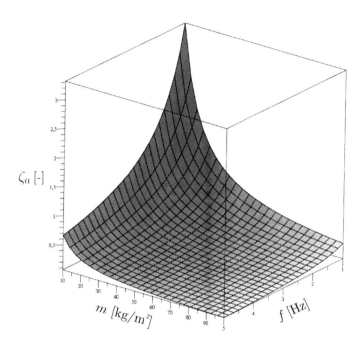

Figure 2: Aerodynamic damping ratio $\zeta_a$ for $c_D = 1$ vs. mass $m$ and eigen frequency $f$.

and sound speed and on the structural properties mass and eigen frequency,

$$\zeta_a = \frac{c_D \rho c}{2m\omega} = \frac{c_D \rho c}{4\pi f m}. \tag{3}$$

Figure 2 illustrates the aerodynamic damping ratio $\zeta_a$ over specific mass and eigen frequency. For normal reinforced concrete and steel structures, the aerodynamic damping ratio is negligibly small. However, with decreasing mass and eigen frequency, the aerodynamic damping ratio increases significantly and reaches values up to 300% for very light and flexible systems (e.g. membrane structures, facades or blast curtains) compared to heavy rigid structures.

It may be of interest that this approach is also implemented in the European standard Eurocode 1, part 7, for the dynamic analysis of structures subjected to wind loads. In the Eurocode 1, the specific logarithmic damping decrement is defined as

$$\Lambda_a = \frac{c_D \rho}{2 f m} v_m \tag{4}$$

where $\Lambda_a = 2\pi\zeta_a$, and $v_m$ is the average wind velocity.

# 3 Analytical FSI model (TG-model)

This section briefly describes the incident and reflected blast waves. Then, a new analytical FSI model (TG-model) is derived which is valid for the analysis of structures subjected to low-level blast loads.

## 3.1 Incident blast wave

Generally, the incident blast wave overpressure history $p_{10}(t)$ can be modeled with

$$p_{10}(t) = \hat{p}_{10}\varphi(t) \tag{5}$$

where $\hat{p}_{10}$ is the incident peak overpressure, and $\varphi(t)$ is a shape function describing the decay behaviour. Often, a simple triangular shape function is used, but recent studies [6, 7] have shown that the negative phase can significantly influence the structural response of flexible systems subjected to air blast loads.

Considering the negative phase, the *Friedlander* approach [8, 9]

$$\varphi(t) = \left(1 - \frac{t}{t_d}\right) e^{-\alpha \frac{t}{t_d}} \tag{6}$$

should be used as shape function for the incident blast wave $p_{10}(t)$. To accurately model the negative phase, the shape parameter $\alpha$ can be determined with

$$\alpha = 1.5\, Z^{-0.38} \quad \text{for } 0.1 < Z < 30\,[\text{m/kg}^2] \tag{7}$$

based on the studies conducted by Borgers and Vantomme [10] where $Z = R/M_{TNT}^{1/3}$ is the scaled detonation distance [11]. In comparison to other formula for the shape parameter, e.g. [3, 11], this approach reproduces the overpressure and especially the underpressure phase more accurately.

## 3.2 Reflected blast wave

When a blast wave hits a structure, it is reflected, and the reflected pressure acts on the structure. Generally, coupling effects are neglected, and the reflected pressure-time variation can then be determined with standard procedures, e.g. [2, 3, 11]. However, the deflection of the structure might influence the reflected pressure as shown in [12]. Due to the nonlinear characteristics of shock waves, general analytical solutions considering these coupling effects are not available. However, for the limit case of low-level blast loads, the derivation of an analytical solution is possible (Section 3.3).

The reflection coefficient $c_r$ describes the ratio of the peak reflected and the peak incident overpressure [13],

$$c_r = \frac{\hat{p}_{r0}}{\hat{p}_{10}} = \frac{8\hat{p}_{10} + 14p_0}{\hat{p}_{10} + 7p_0}. \tag{8}$$

For the limit of a very weak shock wave $c_r$ is close to 2 as illustrated in Figure 3. Figure 3 also shows the often used threat levels GSA C and GSA D according

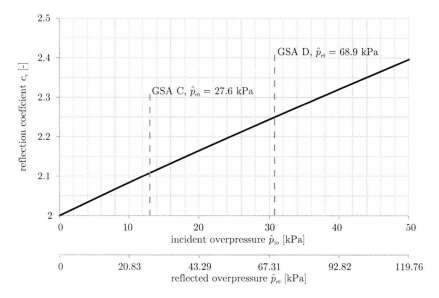

Figure 3: Reflection coefficient $c_r$ vs. incident overpressure $\hat{p}_{10}$.

to the U.S. General Services Administration (GSA) standard [14]. GSA level C is a far field threat scenario with a reflected peak overpressure of 27.6 kPa and a reflected impulse of 193.1 Pa s. This corresponds to a peak incident overpressure of 13.1 kPa and a reflection coefficient of approx. 2.1. This is a typical low-level blast load.

For weak shock waves the conservation equations of mass, momentum and energy can be transformed into a linear wave equation [12, 13]. For this linear model, the principle of superposition is valid, and the reflected pressure history acting on a structure reads

$$p_{r0}^{\text{FSI}}(t) = \hat{p}_{10}\left[\varphi(t) + \phi(t)\right] \tag{9}$$

where $\phi(t)$ is the unknown shape function of the reflected wave, and the index FSI indicates the influence of FSI effects. Neglecting interaction effects yields $\phi(t) = \varphi(t)$, and the reflected pressure-time variation can be written as $p_{r0}(t) = c_r\hat{p}_{10}\varphi(t)$ with $c_r \approx 2$ for very weak shocks.

This linear approximation corresponds to the acoustic limit and is based on two main assumptions: (1) The speed of the blast wave is close to sound speed. (2) The medium transporting the blast wave is almost incompressible. These two assumptions are valid for the propagation of weak and medium shock waves in water. For air blast waves, however, the assumptions are only valid for low-level air blasts with small overpressures and a reflection coefficient of approximately 2.

### 3.3 Equation of motion

As illustrated in Figure 4, the equation of motion for a single degree of freedom (SDOF) model reads

$$m\ddot{x} + d_s + d_a + r = p_{r0}^{\text{FSI}} = \hat{p}_{10}[\varphi + \phi] \tag{10}$$

with the specific mass $m$, the structural damping pressure $d_s = d\dot{x}$, the aerodynamic damping pressure $d_a$ and the resistance function $r$ which depends on the deflection $x$. $p_{r0}^{\text{FSI}}$ is the reflected blast pressure. Both the displacement $x$ and the shape function $\phi$ are unknown. However, the condition of continuity requires that the structural velocity $\dot{x}$ needs to be equal to the sum of the wave velocities of the incident and of the reflected waves [15, 16]. Applying the conservation of momentum, $p = \rho u c$, yields *Taylor*'s FSI relation (see [12, 15] for more details)

$$\dot{x}(t) = \frac{\hat{p}_{10}}{\rho c}[\varphi(t) - \phi(t)] \tag{11}$$

or, equivalently,

$$\phi(t) = \varphi(t) - \frac{\rho c}{\hat{p}_{10}}\dot{x}. \tag{12}$$

This FSI relation couples the structural velocity $\dot{x}$ with the incident and the reflected wave shape functions $\varphi$ and $\phi$. $\rho$ is the density under ambient conditions (for air: $\rho = 1.225 \text{ kg/m}^3$), and $c$ is the sound speed (for air: $c \approx 343$ m/s).

With (12), the reflected pressure history (9) can be written as

$$p_{r0}^{\text{FSI}}(t) = 2\hat{p}_{10}\varphi(t) - \frac{\rho c}{\hat{p}_{10}}\dot{x}. \tag{13}$$

Thus the reflected pressure acting on a structure is altered depending on the structural velocity $\dot{x}$.

Combining (10) and (13) wields

$$\ddot{x} + \left(\frac{d_s}{m} + c_D\frac{\rho c}{m} + \frac{\rho c}{m}\right)\dot{x} + \frac{r}{m} = \frac{2\hat{p}_{10}\varphi}{m} \tag{14}$$

or

$$\ddot{x} + \left(\frac{d_s}{m} + c_D\frac{\rho c}{m} + \frac{\rho c}{m}\right)\dot{x} + \frac{r}{m} = \frac{c_r\hat{p}_{10}\varphi}{m} \tag{15}$$

which is strictly only valid for $c_r = 2.0$. However, we propose $c_r \approx 2.1$ as limit of application. The initial conditions are usually $x(0) = 0$ and $\dot{x}(0) = 0$.

The mass $m$ significantly influences the coupling effects. With increasing mass, $\rho c/m \to 0$ resulting in $p_{r0}^{\text{FSI}}(t) = 2\hat{p}_{10}\varphi(t)$ and $\phi(t) = \varphi(t)$. With decreasing mass and stiffness, however, coupling effects lead to a shape function $\phi(t)$ depending not only on time $t$ but also on the structural velocity. For very light systems, aerodynamic damping and the FSI term $\rho c/m$ dominate the solution of the differential equation (15) and the reflected pressure history $p_{r0}^{\text{FSI}}(t)$ (13). The vibration of the SDOF system is damped by three effects: (1) structural damping $d_s$, (2) aerodynamic damping $d_a$, and (3) damping due to FSI effects (term $\rho c/m$). However, only structural damping leads to internal forces in the system.

## 4 Examples

To illustrate the coupling effects and their implication for practical design applications, we now turn to some numerical examples and analyze several systems with different eigen frequencies and masses subjected to a blast load as a result of the detonation of 100 kg TNT equivalent at a stand-off distance of 40 m. This blast load corresponds approximately to the threat level C according to the U.S. GSA standard [14] and to the threat level EXV 33 according to the European standard [17]. The blast load parameters are summarized in Table 1. The incident peak pressure and incident impulse are computed with the formula of Kinney and Graham [11].

Generalized single degree of freedom (SDOF) models are a cost-effective alternative to powerful, but expensive, FE models. They are extremely useful for preliminary calculations and to check the results of more complex FE analyses. In contrast to FE modeling, they require very limited input data and are suitable for rapid analysis to get an "engineer feeling" for the structural performance. The approach to derive generalized SDOF models of real structures is well established in structural dynamics [18, 19] and is not repeated in this paper.

This section concentrates on a protective design in the linear elastic range where the resistance function reads $r = kx$. The circular eigen frequency is

Table 1: Blast load scenario: spherical detonation of 100 kg TNT equivalent at a stand-off distance of 40 m.

|  | SYMBOL | VALUE | UNIT |
|---|---|---|---|
| scaled distance | $Z$ | 8.62 | m/kg$^{1/3}$ |
| incident peak pressure | $\hat{p}_{10}$ | 12.2 | kPa |
| incident impulse | $i^+$ | 113.1 | kPa ms |
| overpressure time duration | $t_d$ | 22.8 | ms |
| reflection coefficient | $c_r$ | 2.1 | - |
| shape parameter | $\alpha$ | 0.66 | - |

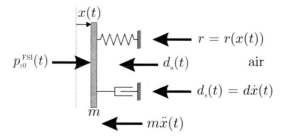

Figure 4: SDOF model: equilibrium of forces.

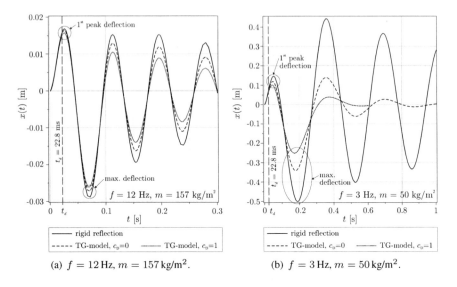

(a) $f = 12\,\mathrm{Hz}$, $m = 157\,\mathrm{kg/m^2}$.

(b) $f = 3\,\mathrm{Hz}$, $m = 50\,\mathrm{kg/m^2}$.

Figure 5: Displacements $x(t)$ over time for two systems with different eigen frequencies and different masses subjected to a detonation of 100 kg TNT equivalent at a stand-off distance of 40 m (3% structural damping).

$\omega = \sqrt{k/m} = 2\pi f$ with the eigen frequency $f$. The differential equation reads

$$\ddot{x} + \left(\frac{d_s}{m} + c_D\frac{\rho c}{m} + \frac{\rho c}{m}\right)\dot{x} + \omega^2 x = \frac{c_r\hat{p}_{10}\varphi}{m} \tag{16}$$

with structural damping $d_s = d\dot{x} = 2m\omega\zeta_s\dot{x}$.

In order to illustrate the basic effects of FSI and aerodynamic damping, Figure 5 shows the time dependant displacements of two systems. System 1 is moderately stiff ($f = 12\,\mathrm{Hz}$) with a specific mass of $m = 157\,\mathrm{kg/m^2}$. These properties correspond to a 2 cm thick, one-way steel plate with the dimensions 2 m x 2 m. The second system is more flexible ($f = 3\,\mathrm{Hz}$) and has a specific mass of $m = 50\,\mathrm{kg/m^2}$ like typical facade and glazing systems.

Figure 5(a) compares the displacements of the first moderately stiff system when assuming rigid reflection (which is the typical approach in protective design) and when considering coupling effects based on the derived TG-model. The coupling effects are divided into two parts: (1) purely FSI without aerodynamic damping ($c_D = 0$) and (2) FSI and full aerodynamic damping ($c_D = 1$).

The first peak displacement of the moderately stiff and fully coupled system (dotted line in Figure 5(a)) is about 9% smaller than the first peak displacement of the rigid analysis (solid line). As expected, the deformation attenuate quickest for the fully damped model. The deflection curves also underline the significance of the negative phase. For flexible systems, the negative phase usually determines the maximum deflection which is then directed contrary to the initial loading direction [7].

For the second and more flexible system shown in Figure 5(b), the coupling effects reduce the first peak displacement by approximately 40% and result in an even quicker attenuation of the time-dependant deformations. The coupling effects reduce the overpressure time duration of the reflected pressure history and the negative phase starts earlier. This contributes to the significant reduction of the first peak displacement.

All in all, the more flexible and the lighter the system is, the more significant are the coupling effects due to FSI and aerodynamic damping. There is a significant energy exchange between the fluid and the structural system.

The solutions for the shape functions $\phi(t)$ are shown in Figures 6 (a)–(b) for the flexible system with $f = 3\,\text{Hz}$ when (a) considering or (b) neglecting aerodynamic damping effects. Figures 6 (c)–(d) show the corresponding results for the stiffer system with $f = 12\,\text{Hz}$. Thus, Figures 6 (b) and (d) show the reflected wave profile $\phi(t)$ when purely considering FSI effects while Figures 6 (a) and (c) show $\phi(t)$ when taking account of FSI and full aerodynamic damping. The drag coefficient $c_D$ controls the amount of aerodynamic damping and is expected to be $0 < c_D < 1$ depending on shape and geometry. The value $c_D$ can only be obtained experimentally.

Figure 6 shows how the coupling effects due to FSI and aerodynamic damping influence the reflected wave profile $\phi(t)$ and thus the reflected pressure-time variation $p_{r0}^{\text{FSI}}(t)$. The smaller the mass, the smaller the overpressure time duration becomes. Thus, the positive impulse is also smaller for light structures. This is especially true for flexible systems with small eigen frequencies.

The coupling effects influence the negative phase, too. For the flexible system in Figure 6 (a), the negative impulse is also smaller than the impulse of the incident wave profile $\varphi(t)$. Without aerodynamic damping, there might be a second positive impulse after the negative phase as shown in Figure 6 (b) for the flexible system. Especially for very light systems, e.g. membrane systems or cable net facades, aerodynamic damping should be taken into account. For heavier systems with specific masses of $m \gtrsim 100\,\text{kg/m}^2$, the influence of aerodynamic damping is very small since $\zeta_a \to 0$.

Summarizing, we state that mass and stiffness influence the reflected wave profile $\phi(t)$. For light and flexible systems with low eigen frequencies, the overpressure time duration, the positive and the negative impulse decrease with decreasing mass. The influence becomes very significant for flexible systems with a specific mass $m \lesssim 50\,\text{kg/m}^2$. These light and flexible systems elude the applied blast load by undergoing large deflections.

## 5 Summary and conclusions

Relatively simple analytical models are important for the validation of complex numerical models and to give the engineer a feeling for the main parameters influencing the numerical results.

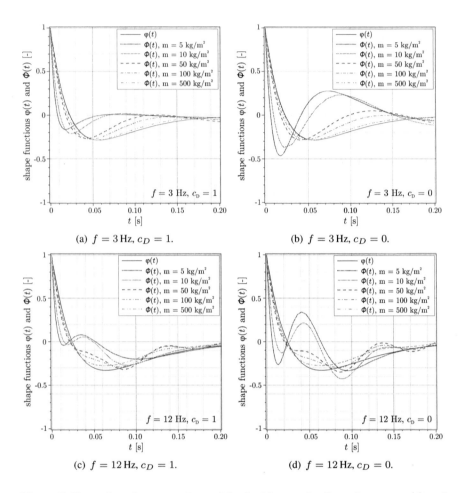

Figure 6: Shape functions over time of the incident and reflected waves, $\varphi(t)$ and $\phi(t)$, for different specific masses $m$ considering ($c_D = 1$) or neglecting ($c_D = 0$) aerodynamic damping. The blast scenario is a detonation of 100 kg TNT equivalent at a stand-off distance of 40 m.

The derived linear TG-model is such a basic SDOF model and valid for the analysis of weak shocks with a reflection coefficient of approximately 2. The authors propose $c_r < 2.1$ as limit of application for the TG-model. This corresponds to a blast scenario with a scaled distance $Z > 8.6 \, \mathrm{kg/m^{1/3}}$ and an incident peak overpressure $\hat{p}_{10} < 12.3 \, \mathrm{kPa}$, respectively (approx. GSA threat level C [14], or European threat level ISO EXV 33 [17]). For stronger blast loads, compressibility effects become important which are not included in the presented TG-model. However, for low-amplitude shocks the TG-model might contribute to a deeper understanding of the physical coupling aspects.

The numerical studies highlight the effects of FSI and aerodynamic damping on the displacements and on the reflected shape function. For flexible systems with low eigen frequencies, the overpressure time duration, the positive and the negative impulse decrease with decreasing mass. The influence becomes very significant for flexible systems with a specific mass of $m < 50\,\text{kg/m}^2$. The coupling effects largely influence the attenuation behavior.

## References

[1] National Research Council (NRC), *Protecting Buildings from Bomb Damage. Transfer of Blast-Effects Mitigation Technologies from Military to Civilian Applications.* National Academy Press: Washington, D.C., 1995.

[2] US Army Corps of Engineers, *TM 5-1300: Structures to Resist the Effects of Accidental Explosions*, 1990.

[3] The Defence Special Weapons Agency and the Departments of the Army, Air Force, and Navy, *TM 5-855-1: Design and Analysis of Hardened Structures to Conventional Weapons Effects*, 1997 *(for official use only)*.

[4] Vickery, B.J. & Kao, K.H., Drag or along-wind response of slender structures. *Journal of the Structural Division*, **98**, pp. 21–36, 1972.

[5] Rickman, D.D. & Murrell, D.W., Development of an improved methodology for predicting airblast pressure relief on a directly loaded wall. *Journal of Pressure Vessel Technology*, **129(1)**, pp. 195–204, 2007.

[6] Krauthammer, T. & Altenberg, A., Negative phase blast effects on glass panels. *International Journal of Impact Engineering*, **24(1)**, pp. 1–17, 2000.

[7] Teich, M. & Gebbeken, N., The Influence of the Underpressure Phase on the Dynamic Response of Structures Subjected to Blast Loads. *International Journal of Protective Structures*, **1(2)**, pp. 219–233, 2010.

[8] Bulson, P., *Explosive Loading of Engineering Structures. A History of Research and a Review of Recent Developments.* E&F Spon: London, 1997.

[9] Friedlander, F.G., Note on the diffraction of blast waves by a wall. UK Home Office Dept, RC(A), 1939.

[10] Borgers, J. & Vantomme, J., Improving the accuracy of blast parameters using a new Friedlander curvature $\alpha$. *DoD Explosives Safety Seminar*, Palm Springs, CA, 2008.

[11] Kinney, G.F. & Graham, K.J., *Explosive Shocks in Air.* Springer: New York, 1985.

[12] Teich, M. & Gebbeken, N., Interaction of air blast waves with light and flexible structures. *Design and Analysis of Protective Structures (DAPS)*, Singapore, 2010.

[13] Courant, R. & Friedrichs, K.O., *Supersonic Flow and Shock Waves*, volume 21 of *Applied Mathematical Sciences*. Springer-Verlag: New York, 1976. (reprint of the edition of 1948).

[14] US General Services Administration, GSA-TS01-2003: Standard Test Method for Glazing and Window Systems, Subject to Dynamic Overpressure Loadings, 2003.

[15] Taylor, G.I., The pressure and impulse of submarine explosion waves on plates. *The Scientific Papers of Sir Geoffrey Ingram Taylor*, ed. G.K. Batchelor, Cambridge at the University Press: Cambridge, volume III – Aerodynamics and the Mechanics of Projectiles and Explosions, chapter 31, pp. 287–303, 1963.

[16] Kriegsmann, G.A. & Scandrett, C.L., Numerical studies of acoustic pulse scattering by baffled two-dimensional membranes. *Journal of the Acoustical Society of America*, **79(1)**, pp. 9–17, 1986.

[17] International Organization for Standardization, ISO 16933:2007: Glass in building – Explosion-resistant security glazing – Test and classification for arena air-blast loading, 2007.

[18] Biggs, J.M., *Introduction to Structural Dynamics*. McGraw-Hill: New York, 1964.

[19] Krauthammer, T., *Modern Protective Structures*. CRC, 2008.

# Effect of silty-sand compressibility on transferred velocity from impulsive blast loading

K. Scherbatiuk[1], D. Pope[2], J. Fowler[1] & J. Fang[3]
[1]*DRDC Suffield, Canada*
[2]*DSTL Porton Down, UK*
[3]*Amtech Aeronautical, Canada*

## Abstract

A common assumption made in calculating response of walls subjected to blast loading is that the reflected blast loading is often based on rigid body dynamics along the thickness of walls and zero particle velocity at the fluid-structure interface. Of interest to the authors is the development of a quick-running impulse-dominated structural response model for soil-filled concertainer walls that assumes reflected impulse assuming zero particle velocity during the course of loading. The magnitude of response will directly depend on the velocity transferred in the thickness direction by the blast loading. The aim of this study is to assess whether a reduction factor should be applied to reduce the initial velocity corresponding to the zero-particle-velocity reflected impulse for concertainer walls filled with compressible silty-sand. Results obtained from a 1-D numerical program using a simple analytical coupling model are validated with those obtained from a commercial coupled Computation Fluid Dynamics and Computational Solid Mechanics (CFD/CSM) code. Both coupled and uncoupled loading combinations are investigated for rigid body, elastic, and compressible silty-sand material behaviours using the simple 1-D numerical program developed and validated. Graphical results for a range of peak blast pressures are compiled to present the differences in zero-particle-velocity reflected impulse required to attain the same velocity in the thickness direction for each different combination. The resulting differences in charge standoffs between each combination are compared as well for a number of different charge sizes. For this specific problem of calculating transferred velocity in the thickness direction for a silty-sand filled concertainer wall subjected to impulsive blast loading, results that include coupling only lead to a small difference in the resulting charge standoff and therefore reasonable results will be achieved assuming uncoupled loading and rigid body dynamics.
*Keywords: blast, impulse, soil, sand, rigid-body, compressibility, concertainer, Hesco.*

# 1  Introduction

Defence Research and Development Canada – Suffield is currently undertaking a four year research program which involves the recommendation of expeditionary structures that exhibit improved performance and protection against blast loading. Soil-filled welded-wire-mesh geotextile-lined concertainers are frequently used to construct walls for defensive structures such as the bunker shown in fig. 1. Of interest is the ability to predict the final response of these structures subjected to blast loading.

Accurate results have been achieved by Pope [1] by utilizing a fully coupled commercial CFD/CSM code. Fully coupled calculations involve computing the effect of the loading on the structure and computing the effect of the structures movement on the loading simultaneously. Uncoupled calculations involve computing the loading based solely on the initial position of the structure and then applying the loading to the structure face assuming that the loading does not change regardless of movement at the fluid-structure interface. However computational effort for both coupled and uncoupled calculations can be extensive for physically large structures that are impulse-dominated in their response. Often the size of elements that are used within a computational model for physically large structures cannot be scaled correspondingly larger as well. Thus large numbers of elements are often necessary for physically large structures while constraints in time step size must still be maintained. Impulsively dominated structures also respond very slowly compared to the relative time duration of a typical blast loading, contributing additionally to long computational run times. Hence a need exists for a quick-running concertainer wall response model that yields reasonably accurate results.

Figure 1:     A bunker constructed using soil-filled concertainers.

Detailed fully coupled fluid-structure models for soil-filled concertainer walls using a commercial coupled CFD/CSM code have been constructed by Pope [1] and experimentally validated by Fowler [2]. Shock tube experiments by Murray *et al.* [3] have been conducted to investigate velocity profiles and shock trajectories attained from a blast wave impacting rigid movable wall. In this study, the accuracy of an analytical solution by Meyer [4] computing the velocity-time history of a rigid movable wall impacted by a shock wave was investigated and found to be in good agreement for weak shock waves, but underestimated reflected loading for strong shock waves. Impulse reduction coefficients have also been investigated by Szuladzinski [5] who studied the effect of varying geometries and mass densities of cross sections on transferred velocity from impulsive blast loading using a commercial coupled CFD/CSM code.

The aim of this paper is to compute, investigate, and compare the magnitude of differences in charge standoffs required for the same velocity in the thickness direction to be attained between uncoupled calculations and coupled calculations for the material types of rigid, elastic, and compressible silty sand. The magnitude of the differences in results will determine if an impulse reduction factor is warranted as a component in the formulation of an analytical/semi-empirical impulse-dominated quick-running blast response model for soil-filled concertainer walls.

To undertake this study, the response a CFD/CSM model of a simple soil-filled concertainer wall is studied in terms of spatially averaged velocity transferred in the thickness direction from an impulsive blast loading. A simple 1-D finite element model is formulated representing a core of soil through the wall thickness using equations from Meyer [4] to couple the blast loading. An equivalent 1-D model is constructed using a commercial coupled CFD/CSM code. The results from the proposed simple 1-D model will be compared and validated with results from the commercial CFD/CSM model in terms of spatially averaged velocity-time histories. An automated root solving procedure that calls the simple 1-D model as a program function will then be derived, coded and developed to solve the inverse problem of calculating zero-particle-velocity reflected pressure and impulse iso-velocity lines for various combinations of uncoupled versus coupled loading, rigid body, elastic, and compressible silty-sand material behaviours. The results plotted graphically on the zero-particle-velocity reflected pressure and impulse plane along with the performance of a range of charge sizes as a function of standoff show the differences in results between uncoupled and coupled loading for the different material behaviours. Magnitudes of differences in standoffs arising from differences in behaviour of the various materials and uncoupled versus coupled combinations will determine whether a reduction factor should be applied to zero-particle-velocity reflected impulses to result in accurate transferred velocities in the thickness direction.

## 2 Description of response of concertainer walls

A model of a simple soil-filled concertainer wall subjected to uniform blast loading was constructed using a commercial CFD/CSM code. Fig. 2 compares the wall's initial position, its position immediately following the blast loading at a time of 30 ms, and its final position at 700 ms.

The strip outlined in the wall's initial position represents a typical core of soil through the thickness of the wall that is assumed to undergo uniaxial strain during the course of the blast loading. This assumption is made with the exception of the areas of the wall that are too close to any edges to strain uniaxially. Comparison of the initial wall position to its position after the duration of the blast loading indicates that the wall has not attained appreciable deformation at this point. Yet detailed modelling results show that the wall has attained a spatially averaged velocity in the thickness direction of approximately 2 m/s. The magnitude of deformation of the wall shown in its final position is mostly a result of the momentum imparted to the wall by the blast loading and the subsequent strains and rotations that occur to absorb this momentum. The large differences between the final position and the position after the duration of the blast loading indicate that the response of soil-filled concertainer walls is mostly impulse-dominated.

Figure 2:     Initial position of wall, position after duration of blast load at 30 ms, and final position at 700 ms.

## 3 Validation of a one-dimensional simple coupled model with a detailed coupled CFD/CSM model

A detailed 1-D shock tube model was constructed using a commercial CFD/CSM code to simulate an incident blast wave impacting a frictionless elastic solid under uniaxial strain and monitor its spatially-averaged velocity-time history. A schematic of this model is shown in fig. 3.

A pressure and temperature of 1 atmosphere and 288 K, respectively, were assumed for the air at ambient conditions while a pressure and temperature of 25 atmospheres and 705.81 K, respectively, were assumed for the air at pressurized conditions.  An ideal gas equation of state was used assuming $\gamma$ equal to 1.4. The shock tube extends in both directions for the distances specified to reduce occurrence multiple shock wave reflections, permit the formation of a slight transmitted shock at the opposing end, and to allow for movement of the elastic solid. The density of the elastic solid was assumed to be 1925 kg/m$^3$. The bulk modulus and Poisson's ratio were assumed as 247.6 MPa and 0.3, respectively, to amount to a uniaxial confined effective modulus of 400 MPa.

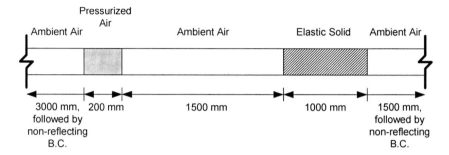

Figure 3:     Detailed shock tube model constructed using a commercial CFD/CSM code.

A simple 1-D numerical model, shown in fig. 4, was formulated using one-dimensional finite elements and equations from Meyer [4] to couple the reflected blast loading.

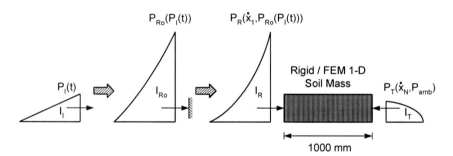

Figure 4:     Schematic of simple model.

The parameter $P_I$ represents the incident pressure time history while the parameter $P_{Ro}$ represents the reflected pressure time history assuming zero particle velocity. The parameter $P_R$ represents the actual reflected pressure-time history applied to the fluid-structure interface and is a function of the velocity and temperature at this interface. Coupling on both the shocked side and the opposing side were included into the model. The pressure on the opposing side is represented by $P_T$, which is the pressure time history at the right side of the mass, and is a function of the velocity at the fluid-structure interface on the right hand side, and the ambient pressure and temperature. The difference equations were derived, coded, and numerically integrated. The displacement at each time step was solved for using fixed-point iteration. The parameters $I_I$, $I_{Ro}$, $I_R$, and $I_T$

represent the impulses corresponding to the pressures described. The reflected pressure-time histories and corresponding temperature distributions corresponding to zero-particle-velocity were calculated using the ideal gas equations from Henrych [6].

Due to difficulties in mitigating the occurrence of multiple reflections in the detailed CFD/CSM model, the zero-particle-velocity reflected pressure and temperature profile were used as inputs into the simple model instead of the incident pressure and temperature profiles, contrary to what is shown in fig. 4. These profiles were computed using the detailed CFD/CSM model by placing a rigid boundary at the location of the left hand side of the elastic solid. The rigid boundary was removed from the detailed model and elements for the elastic solid and additional fluid elements with ambient air were added on the right had side. Both models contain a hundred elements along the thickness of the elastic solid. The velocity-time histories at each element node of the elastic solid in both models were recorded and spatial averages of these velocities at each increment in time were computed.  Fig. 5 compares the spatially averaged velocities computed by both models along with the theoretical velocity-time history calculated assuming uncoupled loading and rigid body dynamics.

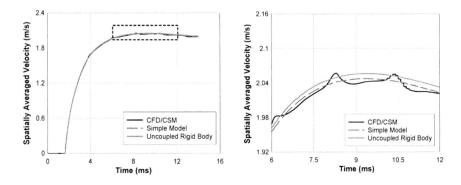

Figure 5:    Comparison of spatially averaged velocities for (a) entire profile, and (b) magnification of profile contained within dotted box.

On average, reasonable agreement is achieved between the detailed CFD/CSM model and the simple model, both attaining average velocities short of the velocities calculated assuming uncoupled loading and rigid body dynamics. The author is uncertain why the spatially averaged velocities computed using the detailed CFD/CSM model appear to oscillate, although the timing of the oscillations do appear to correspond with the time it takes for a compressive wave to travel from one side to the other.

The simple model was created to expedite the computing of velocity results, and to provide well-defined distributions of pressure waves of a specified incident triangular shape, zero time to rise, and absence of a negative phase so that compounded effects from variables not intended to be examined in this study could be eliminated. Although some of these real effects may be valuable to examine, the complexity in the presentation of the results are greatly increased with even one added dimension created by an additional variable.

## 4   Analysis and discussion of differences in coupled and uncoupled results

In developing a quick-running, impulse-dominated blast response model, it is important to quantify what differences exist in transferred velocities between the cases of uncoupled versus coupled loading and their corresponding effects on charge standoffs. The following analysis is aimed to quantify the differences in charge standoffs required to produce identical velocity results if one assumes a rigid material along the thickness, versus an elastic material or compressible silty-sand. Thus the six different cases that are compared are uncoupled-rigid, uncoupled-elastic, uncoupled-compressible, coupled-rigid, coupled-elastic, and coupled-compressible.

The uniaxial stress-strain relationship assumed for a compressible silty-sand is shown in fig. 6. This uniaxial stress-strain relationship and the material coefficients were approximated based on engineering judgement of actual uniaxial compression test results within the MPQW soils database [7]. The effective uniaxial moduli $E_1$ and $E_2$ were selected as 0.4 GPa and 100 GPa respectively. The limiting strain $\varepsilon_a$ was selected as 25%. For tensile strains, a uniaxial effective modulus equal to $E_1$ was selected and the maximum cohesive tensile stress limit was selected as 50 kPa. The density of the soil for all models as well as the rigid body calculations was selected as 1925 kg/m$^3$. For calculations involving an elastic solid material type, effective uniaxial modulus $E_1$ was assumed. The thickness of the soil mass was assumed as 1m for the entire analysis.

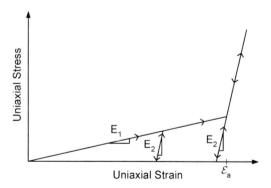

Figure 6:     Uniaxial stress-strain relationship for silty-sand.

There are a number of possible ways one may study the differences in transferred velocity between the six cases. The easiest way would be to use the same incident loading profile and compare differences in transferred velocities for each case. However it is likely that a quick-running model would utilize zero-particle-velocity peak reflected pressure and positive impulse calculated from simple blast loading reflection models as an input. This warrants a comparison of the differences in the zero-particle-velocity reflected impulse necessary to accelerate all six cases to equal velocities in the thickness direction. This is more computationally demanding and requires a root solving scheme to find the solution to this inverse problem. But the variation of results can be easily and directly compared to charge performance with respect to variations in standoffs.

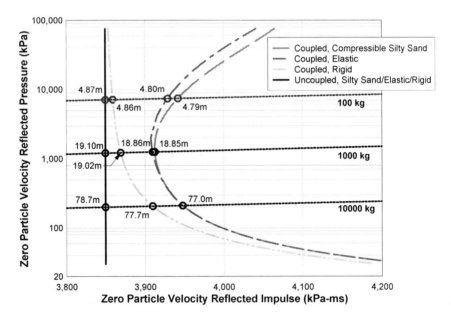

Figure 7:    Zero-particle-velocity reflected impulse and pressures required to accelerate a 1 m thick mass to a velocity of 2 m/s in the six different calculation cases.

Figure 7 shows the results specified in required zero-particle-velocity peak reflected pressure and impulse to accelerate the six different cases to a velocity of 2 m/s. The simple 1-D numerical model described in the previous section was utilized along with an automated root-solving procedure to compute these results. The coupling was turned off in the numerical program to compute the results involving uncoupled calculations. Calculations were carried out assuming a triangular shaped incident blast wave up to the duration time of the blast loading and the spatially averaged velocity was noted at this point. If the peak incident pressure and incident positive impulse selected by the root solver yielded a spatially-averaged velocity of 2 m/s within the tolerances of the root solver, the zero-particle-velocity reflected pressure and impulse were calculated for the corresponding incident blast loading and were plotted on the graph to generate the curves shown. Parametric curves in standoff for zero-particle-velocity reflected impulse and pressure for charge sizes of 100 kg, 1000 kg, and 10000 kg of TNT in a hemispherical configuration were produced using Blast Effects Computer version 5 [8] and imposed onto the graph.

Note that the magnitude of differences between the lines representing the different cases appears visually exaggerated due to the choice of scale range on the ordinate axis. Also for peak reflected pressures above 5000 kPa, the ideal gas laws and assumptions made in Meyer [4] used to compute the coupling begin to lose validity. The true magnitudes of differences in the lines at higher peak pressures may be somewhat larger but in absence of a more robust model, the author felt it was important to include these ranges of computations to show the general shape of the curves and how they deviate from one another.

All three uncoupled cases of rigid, elastic, and compressible, generated identical results. Thus it can be concluded that, for an uncoupled calculation, since the loading is unaffected by the movement at the fluid-structure interface, regardless of the amount of internal deformation that occurs along the thickness, a specified reflected impulse will always yield the same spatially-averaged velocity. The additional strain energy taken up by the internal deformation is balanced with the additional applied work of the loading moving through the range of motion at the fluid-structure interface. These results agree with the equation $v = I_{Ro}/m$, where $v$ is the average velocity in the thickness direction, $I_{Ro}$ is the zero-particle-velocity reflected impulse, and $m$ is the mass of the thickness of material. The numerical program was checked to see that this output was reached when the coupling was turned off and a rigid mass was used instead of a compressible one. The results were in perfect agreement within the numerical tolerances of the root solver.

At very high peak pressures, the coupled-rigid line approaches the uncoupled-rigid line. This is because the higher the particle velocity of the blast wave, the less significance the comparatively low particle velocity at the fluid-structure interface affects the relief of the loading. For lower reflected pressures with correspondingly lower particle velocities, the particle velocity at the fluid-structure interface becomes comparatively more significant and has increased effect. Observing the lower right-hand-side portion of the coupled-rigid, coupled-elastic, and coupled-compressible lines, these lines will eventually approach a horizontal asymptote. As impulse increases to infinity, these loadings approach ideal step shock waves. In theory, an abject that is hit by an ideal step shock wave of infinite impulse will not approach infinite velocity, but approach a maximum velocity limited to the particle velocity of the incident step shock wave.

On the lower right-hand-side of the graph, the coupled-elastic and coupled-compressible lines approach the coupled-rigid line because the pressures are not of a high enough magnitude to cause much internal deformation. Therefore the behaviour of the coupled-elastic and coupled-compressible lines approach the coupled-rigid line for larger charges at further standoffs. For higher pressure, short duration loadings resulting from smaller charge sizes at closer standoffs, the contribution of rigid body velocity becomes very small while the contribution of material deformation increases. Internal deformation is much larger, which causes more movement at the fluid-structure interface and more relief of the loading. Thus more deviation between the coupled-elastic and coupled-compressible lines is seen when compared to the coupled-rigid line. A full discussion of energy losses is beyond the scope of this paper, but the source of these differences more come from relief to the blast loading rather than energy lost due to strain energy absorption. Since all of the uncoupled results regardless of internal deformation received the same applied impulse at the fluid-structure interface, than it follows that in the coupled calculations, although the incident loadings and corresponding zero-particle velocity reflected impulses may be different in all cases, the fluid-structure interface in all cases received an identical amount of applied impulse. Tracking the actual applied impulse at the interface in the coupled calculations validated this. It was the material deformation that caused differences in the actual applied loading.

The differences in standoffs produced by this coupling effect are demonstrated by calculating at what standoffs lines of charge performance intersect the equal-velocity lines for the six cases. For this specific problem, the highest difference produced by this effect was the discrepancy between the 78.7 m standoff and the 77.0 m standoff for the 10000 kg charge, which amounts to a negative 2.16% difference. To summarize, the magnitude of differences in standoffs between uncoupled and coupled calculations for this specific problem of transferred velocity to soil-filled concertainer walls are not large. Therefore the need to account for this behaviour in a quick-running model is small and reasonable results will be achieved if an initial velocity in the thickness direction is calculated using the zero-particle velocity reflected impulse and ignoring the effects of coupling.

## 5 Conclusions

The aim of this study was to assess whether reduction in the transferred velocity corresponding to the zero-particle-velocity reflected impulse is necessary for concertainer walls filled with compressible silty-sand. A quick-running 1-D model using equations in Meyer [4] to couple the blast loading was formulated. The results were compared to results from a fully coupled commercial CFD/CSM code and appear to be in reasonable agreement. The six cases of uncoupled-rigid, uncoupled-elastic, uncoupled-compressible, coupled-rigid, coupled-elastic, and coupled-compressible were investigated. Regardless of the amount of material deformation in the thickness direction, an uncoupled calculation will always yield the same velocity in the thickness direction. For this specific problem of transferred velocity in the thickness direction to a soil-filled concertainer wall, coupling does make a clear difference in results. However the magnitude of this effect only contributes to less than a few percent difference in the charge standoff, and therefore is not worth considering in the formulation of a quick-running impulsive dominated model. Using the zero-particle velocity reflected impulse to calculate the initial velocity in the thickness direction is a reasonable approximation.

# References

[1]     Pope D.J., *The Use of Explicit Element Analysis to Simulate the Behaviour of Military Structures Exposed to Far-field and Near-field Blast Loading*, Technical Report, QinetiQ, UK, pp. 1-52, 2004.

[2]     Fowler J. Personal communication, 15 June 2005, Defence Scientist, DRDC Suffield, Medicine Hat, Alberta, Canada.

[3]     Murray, S.B., Zhang, F., Gerrard, K.B., Guillo, P. and Ripley, R., Influence of diaphragm properties on shock wave transmission. *Proceedings of 24th International Symposium on Shock Waves*, Beijing, China, July 11-16, 2004.

[4]     Meyer, R.F., The impact of a shock wave on a movable wall. *Journal of Fluid Mechanics*, **3**, pp. 309-323, 1958.

[5]     Szuladzinski G. Personal communication, 18 March 2005, Principal, Analytical Service Party Ltd., Sydney, Australia.

[6]     Henrych, J., *The Dynamics of Explosions and Its Use*. Elsevier: New York, pp. 159-163, 1979.

[7]     *MPQW v1.0*. U.S. Army Engineer Waterways Experiment Station, Structural Mechanics Division, Vicksburg, US, 1995.

[8]     Swisdak, M. and Ward, J., *DDESB Blast Effects Computer BEC v5.0*. Explosives Safety Board, Department of Defence, 2001.

# Simplified evaluation of a building impacted by a terrorist explosion

D. Makovička[1] & D. Makovička Jr[2]
*[1]Czech Technical University in Prague,*
*Klokner Institute, Czech Republic*
*[2]Static and Dynamic Consulting, Czech Republic*

## Abstract

This paper determines the parameters of the explosion wave excited by a charge planted by terrorists. A suitcase containing an industrial explosive (Danubit I, mass 6.45 kilogram) remotely controlled by mobile telephone was placed in the left-luggage office of a railway station. Certain simplified methods according to various publications, according to our own experimental results and according to 3D computations based on detailed calculation modelling of the interior of the room are compared to determine the explosive effects. Equivalent static analysis was applied to the dynamic response of the structural elements of the room (walls, floor, roof and windows). The damage caused to these structural elements is weighted on the basis of the angle of fracture of the central axis / surface, and on the basis of the limit stress state of these structures.

*Keywords: explosion, room structure, simplified analysis, dynamic response, failure prognosis.*

## 1 Introduction

When a small charge explodes in the internal space of a building, a pressure wave is formed by the explosion that applies a load on the surrounding internal elements of the structure (fig. 1). The pressure effects of even a small charge are usually high, and the primary consequence is that a window or a door structure may be broken and the pressure is released into the surrounding areas. Although the exhaust vents open, the load transmitted to the surrounding walls of the room, and to the ceiling and floors, is quite high, and the corresponding magnitude must first be estimated. This magnitude depends on numerous parameters that have an impact on the load level, and therefore it is appropriate to adopt simplifying assumptions.

The load of the surrounding structure of the room can be determined either by means of relatively accurate calculations which take into account the internal space of the room, the composition of the explosive, and which deal with the interaction of the internal environment (air and combustion product mixture) with the structure of the room itself. Alternatively, simpler approximate procedures can be applied; these procedures are based on determining the parameters of the explosion load in a free space and then approximating them to the load in a semi-enclosed space (after the exhaust vents have opened).

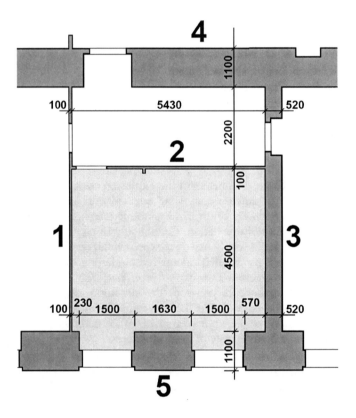

Figure 1:    Ground plan of a room located on the ground floor of a building (the surrounding walls are numbered to provide easier identification of the response calculation).

## 2   Behaviour of masonry when loaded by an explosion

When dynamically loaded by an explosion, masonry displays almost linear elasticity up to the point of failure [6]. The real elastic modulus $E$ is an important material quantity for calculating the response of masonry to the load due to an explosion. It can be estimated according to the deformability modulus $E_{def}$ stated in the design standards, based on the experimentally verified experience of the authors of this paper:

a) To calculate the dynamic response of an undamaged structure near the failure limit:

$$E \approx 0.5 \cdot E_{def} \tag{1}$$

b) To calculate the dynamic response of a structure already damaged by visible cracks:

$$E \approx 0.1 \cdot E_{def} \tag{2}$$

The loading capacity of the brick masonry is decided in accordance with the standards used to design the bending compression / tensile strength of the masonry, with a certain margin of safety, given by the coefficients of the load, combination, etc.

If the loading capacity $R_{tfd}$ is exceeded, a crack appears in the structure of the material. Thus the most unfavourable condition must apply to a safe structure, based on comparing the stress state or the deformation magnitude. The following relationship applies to the stress combination:

$$\min (\sigma_g \pm \sigma_{expl}) \geq -R_{tfd} \text{ or upon adjustment } \sigma_{expl} - \sigma_g \leq R_{tfd} \tag{3}$$

where $\sigma_{expl}$ is stress caused by the effects of the pressure wave when there is an explosion, $\sigma_g$ is the normal stress at a given place (a joint) caused by the overburden weight itself.

In structural design based on limit state theory, it is usually more suitable to consider the carrying capacity strength moment, possibly in combination with the normal force, rather than the carrying capacity limit $R_{tfd}$. This stress criterion must be supplemented by an evaluation of the deformation of the structure. As a rule, the limit deformation value (shift or angular displacement) determines the actual destruction of the wall; the limit deformation value corresponds to the critical angle of the partial turning of the central line of the structure due to its bending. The limit angular displacement $\psi$ at the failure limit is found in the range of approximately 2.3° to 5.7° for masonry [6, 9], a minimum of 6.5° for reinforced concrete [3], and a minimum of 10.5° for steel [3]:

$$\psi = 2 \operatorname{arctg} (2y/l) \tag{4}$$

where $y$ is the maximum achieved deflection of the board (at the centre of the span), and $l$ is the structure span along the shorter dimension.

## 3  Explosion load

When a charge explodes in an open space, the pressure effect of the impact wave on an obstacle (the load of the building structure) depends on the situation of the building with respect to the focus of the explosion, the impact wave parameters, etc. The entire phenomenon of the impact wave effect on the structure is then usually simplified for calculation purposes, using numerous assumptions, especially as regards the intensity and the time course of the impact wave effect and its distribution in contact with the given object [2, 7]. When an actual event takes place, the specific course of the load action depends on the swirl flow bypassing the structure surface, the atmospheric pressure, the temperature conditions and other factors that are usually neglected in a simplified analysis. The parameters of the explosive, too, are determined on the basis of average values; empirical formulas are used, and operate with mean (probable) coefficient levels. Thus the structure calculations concerning the impact wave effects are significantly burdened by these inaccuracies in the input quantities of the entire phenomenon.

Empirical formulas created by various authors [1, 3, 4, 7] are usually used for the time course of the pressure wave and subsequently the structure load. The structure of the formulas according to various authors is very similar, and they usually differ only in the magnitudes of the coefficients. Due to the variability of these coefficients, the uncertainty of the formulas is usually found to be in the range of ±20%, and possibly even more. The reliability of individual formulas improves with increasing distance of the pressure wave from the focus of the explosion.

The overpressure determined at the face of the air impact wave that spreads from the explosion site to the surroundings stems from the reduced distance [1, 3, 7, 8] is:

$$\overline{R} = \frac{R}{\sqrt[3]{C_w}} \tag{5}$$

where $\overline{R}$ is the reduced separation distance from the epicentre of the explosion [m/kg$^{1/3}$], $R$ is the distance from the explosion epicentre [m], and $C_W$ is the equivalent mass of the charge [kg TNT].

It is assumed that the energy released by the explosion is proportional to the mass of the explosive, and the solution consists in introducing a reference charge chosen to be represented by tritol (trinitrotoluene, TNT). Therefore the mass of various explosives is expressed in terms of the so-called tritol equivalent ($k_{TNT}$). If this equivalent cannot be found in the specialized literature (for example [7]), it can be calculated with sufficient accuracy using the relationship

$$k_{TNT\text{-}p} = 0.3\, Q_v - 0.2 \quad \text{(for 2 MJ/kg} \leq Q_v \leq 5 \text{ MJ/kg)} \tag{6}$$

where $k_{TNT\text{-}p}$ is the pressure tritol equivalent of the explosive (equal to 1 for TNT), $Q_v$ is the calculated explosion heat [MJ/kg] and $Q_v = 4.2$ MJ/kg for TNT.

Then the total equivalent mass $C_W$ can be determined using the relationship [7]

$$C_w = C_N \cdot k_{TNT\text{-}p} \cdot k_E \cdot k_G \tag{7}$$

where $C_w$ is the mass of the equivalent charge [kg TNT], $C_N$ is the mass of the used charge of the (actual) explosive [kg], $k_{TNT\text{-}p}$ is the pressure tritol equivalent,

$k_E$ is the charge seal coefficient, and $k_G$ is the geometry coefficient of the impact wave spreading in the space.

The seal coefficient can be determined using the relationship

$$k_E = 0.2 + 0.8 / (1 + k_B) \qquad (8)$$

where $k_B$ is the cover mass [kg] divided by the explosive mass [kg], and expresses the ballistic ratio. The following applies to the geometry coefficient $k_G$, 1 for detonation in a free air space, 2 for surface detonation (on the ground).

The explosion wave spreads in spherical wavefronts from the focus point of the explosion. When the explosion is on the ground, the explosion energy is roughly double because, after complete reflection from the surface of the terrain, the pressure wave spreads in hemispherical wavefronts. The spreading geometry coefficient $k_G$ is not stated by some authors in the formulas for determining the total equivalent mass; in such cases, and in the case of a ground explosion, the equivalent charge mass $C_W$ is as a rule substituted by twice its value in empirical formulas.

In the simplified calculation [7], a ground explosion is represented by a situation when the explosive is located directly on the surface of the terrain ($h = 0$ m thus $k_G = 2$). An explosion in an open air space is a situation when the delay of the reflected wave from the surface of the terrain to the pressure wave front is higher than the duration of the overpressure phase of the pressure wave ($k_G = 1$). A linear interpolation is made between the values.

On the basis of comparing various resources in the literature (namely [1, 3, 4]) and on the basis of tests of bricked structures [2, 6] and window glass [5] during explosions of small charges, the authors of this paper proposed the application of realistic formulas. The empirical formulas below were verified in experiments using small charges (Semtex) in the vicinity of the loaded structure. Their resulting form then corresponds to the impact wave effects from a small solid charge in an external environment and during a ground explosion. Maximum overpressure $p_+$ and underpressure $p_-$ at the face of the air impact wave and their durations $\tau_+$ and $\tau_-$ are applicable both to ground explosions ($C_W$ is replaced by the double value of the equivalent charge) and above-ground explosions in a free (air) environment:

$$p_+ = \frac{1{,}07}{\overline{R}^3} - 0{,}1 \quad \text{[MPa]} \quad \text{for } \overline{R} \le 1 \text{ m/kg}^{1/3} \qquad (9)$$

$$p_+ = \frac{0{,}0932}{\overline{R}} + \frac{0{,}383}{\overline{R}^2} + \frac{1{,}275}{\overline{R}^3} \quad \text{[MPa]} \quad \text{for } 1 < \overline{R} \le 15 \text{ m/kg}^{1/3} \qquad (10)$$

$$p_- = \frac{0{,}035}{\overline{R}} \quad \text{[MPa]} \qquad (11)$$

$$\tau_+ = 1{,}6 \cdot 10^{-3} \cdot \sqrt[6]{C_w} \cdot \sqrt{R} \quad \text{[s]} \qquad (12)$$

$$\tau_- = 1{,}6 \cdot 10^{-2} \cdot \sqrt[3]{C_w} \quad \text{[s]} \qquad (13)$$

After a normal (perpendicular) impact of the explosion wave on a solid obstacle, a reflected wave is formed with the reflection overpressure $p_{ref}$ that loads the building structure from the front side (fig. 2). The overpressure value in the reflected wave corresponds to approximately twice the value of the overpressure for low overpressure values $p_+$ of approximately up to 5 MPa (up to eight times the value for high overpressures of the order of several MPa) in the incident wave for the given distance $R$ [7].

$$p_{ref} \approx 2\,p_+ \tag{14}$$

$$p_{ref} \approx 2\,p \tag{15}$$

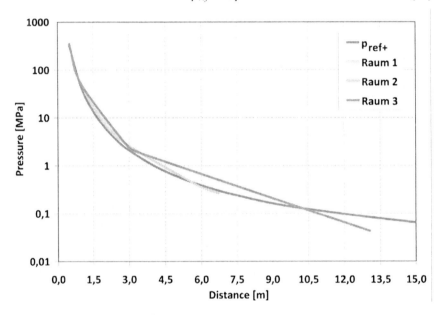

Figure.2:    Magnitude of overpressure $p_{ref}$ in dependence on distance $R$ of the charge location compared to the reflection overpressure $p_{ref}^{f}$ for specific volumes of rooms 1 to 3.

After an explosion in the enclosed space of rooms in a building structure, with closed relieving vents, the load is increased by approximately 50% due to reflection from the surface of the walls, the ceiling and the floor of the room; the duration of the overpressure is thus roughly double. The resulting load of the surrounding structures and its duration can be expressed approximately as follows:

$$p_{load} \approx 1.5 \cdot p_{ref} \tag{16}$$

$$t_{load} \approx 2\,\tau_+ \tag{17}$$

Formulas similar to those for the overpressure phase of the load also apply approximately to the underpressure phase.

The reflective overpressure $p_{ref} = p_{ref+}$ in the rooms can also be calculated directly $p_{ref} = p_{ref}^f$ according to a method described in [9].

To determine the reflective overpressures and impulses, their values in the band $\overline{R} < 2$ m/kg$^{1/3}$ must either be read from the published curves [9], or their approximate values must be determined using the derived exponential relationships:

a) Reflective overpressure:

$$p_{ref}^f = 14,554 \times \overline{R}^{-1,4587} \quad \text{[MPa]} \quad \text{for } 0,05 < \overline{R} \leq 0,5 \text{ m/kg}^{1/3} \quad (18)$$

$$p_{ref}^f = 5,76 \times \overline{R}^{-2,762} \quad \text{[MPa]} \quad \text{for } 0,5 < \overline{R} \leq 5 \text{ m/kg}^{1/3} \quad (19)$$

b) Reflective impulse:

$$I_{ref}^f = 0,345 \times \sqrt[3]{C_W} \times \overline{R}^{-1,857} \quad \text{[kPa.s]} \quad \text{for } 0,05 < \overline{R} \leq 0,5 \text{ m/kg}^{1/3} \quad (20)$$

$$I_{ref}^f = 0,5823 \times \sqrt[3]{C_W} \times \overline{R}^{-1,0976} \quad \text{[kPa.s]} \quad \text{for } 0,5 < \overline{R} \leq 5 \text{ m/kg}^{1/3} \quad (21)$$

When substituting input values $C_W$ into the relationships above, two differences from the calculations of the reflective overpressure and the reflective impulse in the open space outside the building must be taken into account:

a) Here, the indices $f$ denote the detonation conditions in the free air space in the room, and the following is substituted for the charge size $C_W$ (different from formula (3) above):

b) $\qquad\qquad C_w = C_N \cdot k_{TNT \cdot p} \cdot k_E \cdot k_G = C_N \cdot k_{TNT \cdot p} \cdot k_E$ (22)

where the impact wave spreading geometry coefficient $k_G = 1.0$.

## 4   Overpressure calculation for specific rooms

Figure 2 shows the calculation values of the reflection overpressure $p_{ref+}$ calculated using two different simplified procedures, namely according to formulas (9) to (15) derived for an explosion wave spreading in an external space, and furthermore, in order to compare the results of a direct calculation of the reflective overpressure in the rooms, according to formulas (18) and (19) for three selected rooms according to the methodology in [15]:

Room 1 (fig. 1): Volume 69 m$^3$, exhaust vents 1.7 m$^2$, area of the walls, floor and ceiling 104.9 m$^2$;

Room 2: Volume 69 m$^3$, exhaust vents 1.7 m$^2$, area of the walls, floor and ceiling 91.5 m$^2$;

Room 3: Volume 255.8 m$^3$, exhaust vents 24.7 m$^2$, area of the walls, floor and ceiling 283.1 m$^2$.

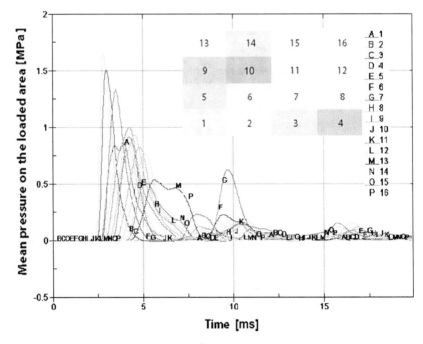

Figure 3:    Time course of the mean pressure on the basis of a 3D calculation of the interaction that occurs between the pressure wave from the explosion of a charge in the centre of the room, and the wall at points corresponding to the wall as divided into rectangles.

A comparison of the results obtained by the two simplified procedures shows that both methodologies provide sufficient accuracy for the usual volume of medium-size and large rooms, and can be applied to any position in which the charge is deposited in the internal areas of the rooms.

Now the calculated overpressures for selected wall 1 (using the marking based on fig. 1) will be compared, as calculated by the LS DYNA program [10], taking into account the interaction of the room environment with its walls (Fluid-Structure Interaction) in fig. 3. The calculated wall is divided into 16 fields, and the mean overpressure for the suitcase containing the explosive, deposited at the centre of the room, is calculated and shown in each of the fields. This figure shows that, due to reflections from the walls of the room, the explosion peaks are superimposed on each other and their coverage curve roughly corresponds to formulas (16) and (17). The peak pressure coverage is dominant for the global structure response. So that an equivalent "static" calculation is justified for the analysis of this problem.

## 5   Calculating the damage to the walls of the rooms

In order to evaluate the masonry of the walls and pillars of the room structure, we can use the load estimate $p_{load+}$ and the duration of its effect $t_{load+}$, calculated for the possible distance $R$ of the charge position from the evaluated wall, window, door or interwindow or inside pillars.

The equivalent static calculation of the wall uniformly and continuously loaded by load $p_{load+}$ with duration of its effect $t_{load+}$ was used to determine the failure hazard. In this procedure, the nature of the boundary conditions (support of the wall board) is chosen. As concerns the partition walls of a specific building, the individual wall board elements were considered to be independent, for the sake of simplification, hinge-mounted along their entire circumference.

From the viewpoint of uncertainty in the simplified effect of the pressure wave from the explosion, even such a simplification is acceptable and justified for an engineering estimate of the explosion effects.

As a rule, the dynamic coefficient $\delta$ is derived for the equivalent static calculation, and for a system with one degree of freedom it is the function of the natural period of dominant oscillation $T$ of the structure and the pressure wave effect duration $\tau_+$ or $\tau_-$, based on whether the overpressure or underpressure wave is considered.

For the elastic-plastic system, dynamic coefficient $\delta$ is the function of the ratio of the impact wave effect duration $\tau_+$ or $\tau_-$ on the natural structure oscillation period $T_{(i)} = T$ and on the ductility of the structure:

$$k_m = \frac{y_m}{y_{el}} \tag{23}$$

where $y_m$ is the total elastic + plastic deflection (shift) of the structure, and $y_{el}$ is the elastic part of the deflection (shift).

As for impact phenomena (very rapid) during bending stress of the structure, the ductility coefficient $k_m$ can usually be considered to be equal to 3 to 5 for masonry, and from 5 to 10 for reinforced concrete, steel and wood. As for the load due to the impact wave, the dynamic coefficient including consideration of the ductile behaviour of the structure is found to be in the range $\delta = 1 \sim 2$. This magnitude was derived by N M Newmark (see [3]) for a simplified system with one degree of freedom in the following form:

$$\frac{1}{\delta} = \frac{T_{(i)} \cdot \sqrt{2 \cdot k_m - 1}}{\pi \cdot \tau_+} + \frac{1 - \dfrac{1}{2 \cdot k_m}}{1 + 0{,}7 \cdot \dfrac{T_{(i)}}{\tau_+}} \tag{24}$$

## 6   Evaluating the failure probability for a specific room

When calculating the load level $p_{load+}$, the load is found to be in the range of units of MPa or hundreds of kPa inside a room (fig. 1), based on the position of the charge inside the room. When comparing such high loads with the carrying capacity of the windows and doors, it reaches several units of kPa. It is apparent that such window and door openings will be smashed (destroyed) and will enable the pressure to be released into the surrounding (external or internal) areas.

Table 1 shows the calculated bending moments in the middle part of the wall board in vertical and horizontal directions, maximum deflection $y$ at the centre of the wall board and the angle $\varphi$ of angular displacement of the centre line of the wall board. The angle of 5° was chosen as the limit angle $\psi$ (table 2) at which the wall board masonry breakdown occurs (fracture, sweeping out of brick fragments, etc.) For the sake of transparency, the individual walls of the room are numbered and these numbers are shown in the ground plan of the room in fig. 1. It follows clearly from table 1 that thin partition walls up to 150 mm in thickness of will be destroyed by the explosion.

Table 1:     Failure risk estimation of structural parts (fig. 1).

| Structural element | Distance of charge | Explosion load | Load duration | Vertical moment | Horizontal moment | Displacement | Rotation | Failure estimation |
|---|---|---|---|---|---|---|---|---|
| | $R$ | $p_{load+}$ | $t_{load+}$ | $M_{ver}$ | $M_{hor}$ | $y$ | $\varphi$ | |
| | [m] | [MPa] | [s] | [kNm] | [kNm] | [mm] | [deg] | |
| | Wall 6700×2800×100 mm | | | | | | | |
| 1 | 1.5 | 20.53 | 0.006 | 360 | 141 | 7088 | 157.7 | Expected |
| | 2.5 | 5.09 | 0.008 | 116 | 45 | 2282 | 117.0 | Expected |
| | Wall 5430×2800×100 mm | | | | | | | |
| 2 | 2 | 9.27 | 0.007 | 188 | 76 | 3697 | 138.5 | Expected |
| | 5 | 0.89 | 0.011 | 28 | 11 | 557 | 43.4 | Expected |
| | Wall 6700×2800×520 mm | | | | | | | |
| 3 | 1 | 64.73 | 0.005 | 4902 | 1924 | 688 | 52.3 | Expected |
| | 3 | 3.15 | 0.009 | 409 | 160 | 57 | 4.7 | Partial failure |
| | 6 | 0.58 | 0.012 | 105 | 41 | 15 | 1.2 | Improbable |
| | Wall 5430×2800×1100 mm | | | | | | | |
| 4 | 1 | 64.74 | 0.005 | 10202 | 4122 | 151 | 12.3 | Probable |
| | 2 | 9.27 | 0.007 | 2 030 | 820 | 30 | 2.5 | Improbable |
| | 4 | 1.52 | 0.010 | 460 | 186 | 7 | 0.6 | Improbable |
| | Pillar 1630×2800×1100 mm | | | | | | | |
| 5 | 1 | 64.74 | 0.005 | 9227 | 4050 | 44 | 6.3 | Partial failure |
| | 2 | 9.27 | 0.007 | 1751 | 769 | 8 | 1.2 | Improbable |
| | 4 | 1.52 | 0.010 | 369 | 162 | 2 | 0.3 | Improbable |

Table 2: Limit failure angle $\psi_{max}$ [°] upon breaking of the material [5, 6, 7].

| Type | Structure material | $\psi_{max}$ [°] |
|---|---|---|
| 1 | Concrete C16/20 to C40/50 | 6.5 |
| 2 | Masonry, full bricks 10, mortar 4 or mortar 10 | 5.0 |
| 3 | Masonry, cement bricks, mortar 4 | 4.5 |
| 4 | Masonry, cellular concrete or perforated precise blocks, mortar 4 | 4.0 |
| 5 | Steel S235 | 10.5 |
| 6 | Wood, hard and soft | 12 |
| 7 | Window glass, thickness 3 mm | 6 |

As the explosion pressures markedly exceed the carrying capacity of such thin partition walls, the ruins of the partition walls will be swept into the surrounding areas. Thick bricked walls and interwindow pillars 900 mm and more in thickness will be destroyed only if the charge is placed in their vicinity, at a distance of about 1 m. For distances of the charge of more than 2 m, such a massive structure will transfer the explosion pressures without collapsing and without any other serious defects. Of course, the plaster will be damaged, cracks will appear in the walls, brick fragments may fall out, etc., but the structure will not collapse.

If a massive carrying wall or pillar (more than 900 mm in thickness) collapses under this ceiling, it is likely that the ceiling structure will fall through and damage will also occur to higher floors.

## 7  Conclusions

An example of a specific building was used to discuss the explosion and the building safety hazard when a terrorist charge is brought into the building in a suitcase and is equipped with a system for initiating the charge after it has been placed in the building and the terrorist has left.

Due to uncertainties in all parameters of the explosion load, a simplified methodology has been presented here. This methodology enables the parameters to be determined sufficiently concisely and the natural building structure to be evaluated on the basis of the parameters. The uncertainty in determining the explosion load parameters can be determined on the basis of the results of a calculation using empirical formulas derived by the authors for small charges. The response of the structure is evaluated on the basis of the results of the equivalent static calculation, using the dynamic coefficient for the elastic-plastic system. The explosion hazard of the structure is evaluated on the basis of the maximum moments and deflections of the structure.

The example of a specific room is used to analyze its exposure and also the hazard to the entire building based on various possible placements of a charge a short distance or a longer distance away from the carrying structure and the partition walls.

## Acknowledgement

This research was supported as a part of the research projects in GAČR 103/08/0859 "Structure response under static and dynamic loads caused by natural and man induced activity", for which the authors would like to thank the Agency.

## References

[1] Henrych, J., *The Dynamics of Explosion and its Use,* Academia: Prague 1979.
[2] Janovský, B., Šelešovský, P., Horkel, J. & Vejs, L., Vented confined explosions in Stramberk experimental mine and AutoReaGas simulation. *Journal of Loss Prevention in the Process Industries,* **19**, pp. 280–287, 2006.
[3] Koloušek, V. et al., *Building Structures under Dynamic Effects* (in Slovak), SVTL: Bratislava 1967.
[4] Korenev, B.G. et al., *Dynamic Calculation of Structures under Special Effects* (in Russian), Strojizdat: Moskva 1981.
[5] Makovička, D., Shock wave load of window glass plate structure and hypothesis of its failure. *Structures under Shock and Impact V,* eds. C.A. Brebbia, N. Jones, G.D. Manolis & D.G. Talaslidis, WIT Press: Southampton, pp. 43–52, 1998.
[6] Makovička, D., Failure of masonry under impact load generated by an explosion. *Acta Polytechnica,* **39(1)**, pp. 63–91, 1999.
[7] Makovička, D. & Janovský, B., *Handbook of Explosion Protection for Buildings* (in Czech), CTU Publishing House in Prague, 2008.
[8] Makovička, D., Makovička, D., Janovský, B. & Adamík, V., Exposure of building structure to charge explosion in interior (in Czech), *Stavební obzor,* **18(9)**, pp. 257–265, 2009.
[9] Baker, W.E., Westine, P.S., Cox, P.A., Kulesz, J.J. & Strehlow, R.A., *Explosion Hazards and Evaluation,* Elsevier: Amsterdam 1983.
[10] LS-DYNA User's Manual, *Nonlinear Dynamic Analysis of Structures,* Version 950, Livermore Software Technology Corporation, May 1999.

# Simplified blast simulation procedure for hazard mitigation planning

T. Tadepalli & C. L. Mullen
*Department of Civil Engineering, University of Mississippi, USA*

## Abstract

A simplified procedure for blast simulation is presented for the purpose of performing building vulnerability assessment in a framework consistent with natural hazards mitigation planning studies. A computational tool, FESIM, has been developed to perform simplified building damage simulation using a GUI driven set of algorithms. FESIM permits estimation of the dynamic response of SDOF and MDOF systems to arbitrary forcing functions, and includes a beam-column damage element for computation of nonlinear response of portal frames including laminated composite beams with varying material and laminate layout sequences. The procedure begins with the formation of stiffness and mass matrices for computation of natural modes and frequencies of the portal frames. Blast loading-time histories are then generated and SDOF elasto-plastic response is computed to determine P-I curves. Geospatial analysis is used to display blast pressure contours for scenario events affecting a complex of buildings. The damage to buildings and debris created are estimated using the P-I curves developed for the class of buildings in the complex, and losses are computed based on the occupancy, structural value and content value. The procedure is demonstrated for buildings located at the University of Mississippi being studied as part of a federally sponsored Disaster Resistant University project.
*Keywords: FE simulation, blast loading, GIS, mitigation planning.*

## 1 Introduction

The design of military structures for blast loads follows well laid design guidelines. The same is not the case when it comes to designing civilian structures for blast resistance. The technology transfer from military to civilian spheres of design has yet not been fully realized. Most of the design and analysis tools are not accessible to civil planners and designers, whose primary focus is on urban structures. It is not feasible to conduct complex computational fluid dynamics (CFD) and finite element (FE) simulations for an urban environment consisting of a very large number of structures. The aim of this work is to provide urban planners, designers and engineers with a simple tool for risk assessment and blast hazard mitigation. The major steps in conventional natural hazards loss estimation methodology are followed in terms of hazard analysis, structural analysis (engineering demand parameters), damage analysis and loss analysis; however, only deterministic scenarios are assessed in this procedure (fig. 1).

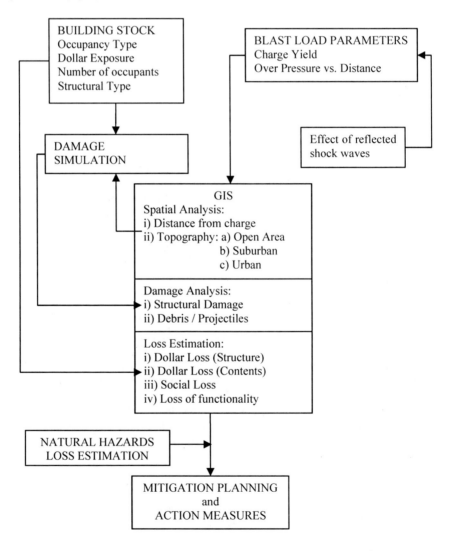

Figure 1:     Proposed blast vulnerability assessment.

## 2   Hazard analysis

The program provides blast overpressure versus distance curves, scaled for various charge yields of TNT based on empirical network type equations [1]. An equivalent triangular load is computed for subjecting the SDOF system to blast loading. For a given charge yield and stand off distance, the peak over pressure and impulse are computed from the scaling law.

The peak overpressures are modified based on the terrain surrounding the building. The contribution of the reflected shocks is determined based on how close the neighbouring structures are to the building(s) of interest. As of now, this is assigned to the structure by defining the terrain as open, suburban (buildings within 100 m of each other) or urban (buildings within 30 m of each other). Later versions of this program will perform spatial analysis to compute distances and orientations between the structures and compute the contributions of the reflected shock waves more accurately.

The amount of debris and fragments generated are computed based on the percentage of glazing on the exposed surfaces, from empirical formulae [1].

## 3   Structural analysis

An interactive computational simulation tool called FESIM which permits an interface between modal vibration analysis testing and finite element (FE) damage prediction has been developed, for composite structures dynamic response analysis. The tool operates in the Matlab (The Mathworks, Inc., Natick, MA) computing environment and is useful for advanced undergraduate/graduate instruction and has been developed in a computational framework that enables a variety of research activities. The FESIM framework includes a variety of time domain based linear and nonlinear dynamic analyses, which can initiate from either experimental modal analysis or FE formulation for structural systems made of advanced composite materials.

The instructional FESIM tool currently provides a 2D frame element for simulating linear static and dynamic analysis of frame structural systems. The research oriented tool is capable of simulating nonlinear dynamic behavior of 3D frame, solid, and shell systems. A graphical user interface (GUI) (fig. 2) has been created to enable simple and direct input by the user including structural property matrices for small instructional problems.

### 3.1 Section tangent stiffness integration and time varying flexural damage simulation in FESIM

For the use of FESIM in an FE environment, a section damage model based on Euler-Bernoulli beam theory has been implemented [2, 3]. Sections are discretized into non-overlapping area patches (constant width rectangles for 2D case) for purposes of integration of stiffness and mass matrices.

A mixed interpolation strategy using linear shape functions for axial response and cubic (Hermitian) shape functions for flexural response is adopted to avoid shear locking that occurs with isoparametric shape functions.

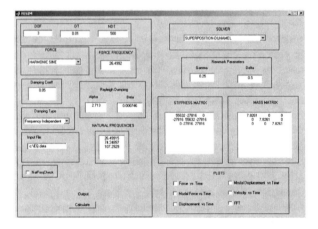

Figure 2:    FESIM GUI shows interactive input and output.

The section stiffness and mass are computed at two Gauss integration points along the length of the element in order to minimize computational effort and storage. The shifting of neutral axis due to damage is also neglected.

The time varying strain versus nodal displacement vector for cross sections located at position $x$ on the neutral axis at time $t$ is obtained from the strain compatibility on the section. The tangent stiffness matrix of the element is obtained by:

$$^t\mathbf{K} = \int_{V^e} \mathbf{B}^T \cdot {}^t D^{NL} \cdot \mathbf{B} \cdot dV \tag{1}$$

where

$$\boldsymbol{B}(x) = \left[ N_{1,x}^L; \, y \cdot N_{1,xx}^H; \, \mathrm{y} \cdot N_{3,xx}^H; N_{2,x}^L; \, y \cdot N_{2,xx}^H; \, \mathrm{y} \cdot N_{4,xx}^H \right]$$

$$^t D^{NL} = {}^t E = \frac{\partial[{}^t\sigma_1]}{\partial[{}^t\varepsilon_1]}$$

The integration in eqn (1) is first performed over the section and then over the length using numerical integration:

$$^t\mathbf{K} = \int_L {}^t\mathbf{K}^A \cdot dx_1 = \sum_{i=1}^{2} {}^t\mathbf{K}^{Ai}(x_i) \cdot w_i \tag{2}$$

where    $^t\mathbf{K}^{Ai} = \int_{Ai} \mathbf{B}^T(x_i) \cdot {}^t E \cdot \mathbf{B}(x_i) \cdot dA$

For the general case where the variation of the tangent modulus $^tE$ with time and location is not defined by an explicit relation, eqn (2) must be integrated numerically. For the $i^{th}$ integration point (section), a typical element of the tangent section stiffness matrix is:

$$'\left[\mathbf{K}_{22}\right]^{Ai} = [N_{1,xx}^{H} \cdot N_{1,xx}^{H} \sum_{j=1}^{na} {}^{t}E_{j} \cdot I_{j}]_{i}$$  (3)

where ${}^{t}E_{ij}$ is the tangent modulus at the centroid of area patch $j$ in section $i$ at time $t$ and

$$I_{j} = I_{j0} + y^{2} \cdot A_{ij}$$

the 2nd moment of area patch $A_{ij}$ about the $z$-axis.

## 3.2  Application to incremental plastic collapse analysis of a cantilever beam

A methodology for nonlinear incremental analysis [4] has been implemented in FESIM using the 2D frame element [3]. At each load increment, the local strain at each area patch centroid is computed, and the corresponding stress is interpolated from the material stress-strain data. This allows for definition of nonlinear material properties for each individual area patch. The tangent modulus is used to compute the stiffness matrix as described above. The modified Newton-Raphson iteration is used to compute the global equilibrium state.

FESIM simulation of the response of a cantilever beam loaded to collapse is performed using a multi-linear, elastoplastic material constitutive law, and the values obtained compare favorably with analysis using ABAQUS (HKS, Inc., Providence, RI) when sufficient numbers of area patches are used.

The section approach used in the stiffness integration allows the damage element to output data at all area patches, which is useful in visualizing the progression of damage both through the thickness and along the length. To illustrate this, a model of the cantilever is used to show the penetration of the elastic-plastic boundary [5] until formation of a plastic hinge at the fixed end (fig. 3). The model only required *two* elements to accomplish the result shown. The routine is also useful in computing the yield displacements of frame structures through a pushover analysis.

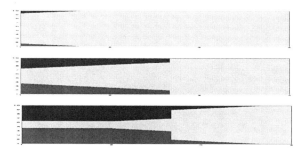

Figure 3:    FESIM output for plastic hinge progression through the length and depth of a cantilever beam.

### 3.3 SDOF elastoplastic time history analysis

The nonlinear dynamic response of elastoplastic systems is computed using the Newmark Method. The response of a SDOF system to a triangular impulse with varying rise times is provided as an example (fig. 4). The peak response is observed to increase with decreasing rise time (tr) for a constant impulse.

Detailed nonlinear dynamic FE modelling of the Student Union building on the Oxford campus of the University of Mississippi was carried out as part of a research project examining seismic vulnerability of campus buildings (fig. 5) [6]. The FE model is used here to develop the lumped parameters for an equivalent SDOF system. At present an SDOF oscillator having yielding bilinear force-displacement response characteristics is utilized for computing pressure-impulse (P-I) curves used in simulating blast response (fig. 4).

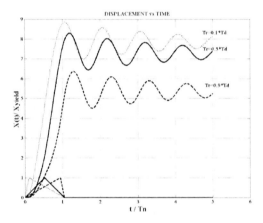

Figure 4:  FESIM output for elasto-plastic SDOF system with varying rise time.

Figure 5:  Detailed nonlinear FEA model of the Student Union Building.

## 4   Pressure-Impulse (P-I) curves

P-I curves are also known as isodamage curves. Each curve represents a certain value for the ductility ratio. For various values of time duration of the impulse ($td$), the time of rise ($tr$), peak load and the impulse are varied. The maximum values of each response are computed, and if they match a certain desired value of the ductility ratio, they are plotted against the normalized peak load and normalized impulse [7, 8].

When a structural type is assigned to a building, pre-computed P-I curves corresponding to that structural type are assigned to the building. Thus, the P-I curves play a role analogous to fragility curves in seismic vulnerability assessment. The P-I curves (fig. 6) for the Student Union building have been computed for ductility ratios of one, two and three.

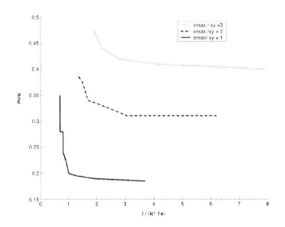

Figure 6:     P-I curves for the Student Union building.

## 5   Geographical Information Systems (GIS)

GIS provides the geospatial analysis for the proposed procedure. The GIS model for blast scenarios (fig. 1) is similar to that used in HAZUS-MH for seismic, wind and flood scenarios. The advantage of geospatial analysis is that the user can see all the processes, relationships and linkages between various parts of the database. A building database may be populated with data obtained from field surveys. Building data relevant to direct physical damage estimation includes structural type, occupancy, number of storeys, glazing (glass, brick veneer) percentage of total exposed building surface area. For socio-economic impact, the dollar replacement value, contents value and the maximum number of occupants associated with each building are included in the building inventory.

A multi-hazard mitigation project [9] sponsored through a nation-wide Disaster Resistant University program is currently underway on the Oxford campus of the University of Mississippi. The proposed blast simulation procedure is applied to a group of the campus buildings to illustrate the potential benefits. A deterministic scenario is considered involving an explosive having a charge yield of 500 kg TNT that is detonated from a vehicle parked on campus. The results and implications are visualized on a GIS map that overlays photos and geospatial relationships of a small group of buildings surrounding the vehicle (fig. 7).

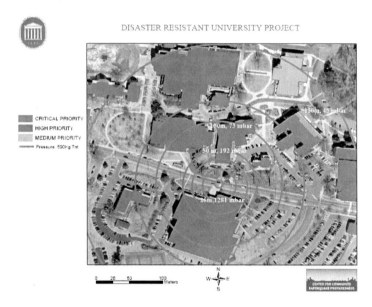

Figure 7:    Hypothetical blast scenario on university campus.

## 6  Damage analysis

Once the correlation between a blast loading event and the damage is established, the response of equivalent SDOF systems is used to develop P-I curves for various structural types. The ratio of peak displacement to the yield displacement of the structure, also known as the maximum displacement ductility demand, is utilized as a parameter for determining the damage to a given building.  For example, in the case of steel structures, a ratio of 3 is considered to correspond to moderate damage [8]. The damage levels are quantified as low, moderate and high. Once the degree of damage in a building due to a blast event is determined, the performance levels of the building are classified into immediate occupancy, life safety and collapse prevention.

## 7  Loss estimation

The loss due to a blast event is computed using a methodology similar to that used for seismic loss estimation. The premise is that the range of damage states which might occur due to a blast is similar to that which might occur during earthquakes. Once the damage due to a blast event is determined, the loss is estimated as a percentage of the building dollar value, the contents value, social loss and loss of functionality. The dollar loss is computed based on the current replacement value, even though the buildings may be constructed in stages. The contents value can be estimated from the building inventory. The average number of people occupying the building at a peak occupancy level is used to determine the social loss. The dollar loss due to non-functionality is computed based on the dollar revenue generated per day by the facility.

Once the risk of a specified level of blast occurring in various locations is estimated and critical facilities are identified based on their economic, administrative, emergency management, operational, and social importance, action measures may be identified that might mitigate the projected losses. Some measures call for more detailed analysis such as hardening of building structures. Other measures relating to policy making and non-intrusive actions such as event control and provision for barriers or buffer zones may be accomplished directly using the simplified procedure.

## References

[1]     Kinney, G.F. and Graham, K.J., *Explosive Shocks in Air*, 2$^{nd}$ ed., Springer Verlag: Berlin, 1985.
[2]     Mullen, C.L. and Cakmak, A.S., A practical damage element for seismic analysis of RC structures, unpublished work in partial fulfilment of the degree of Doctor of Philosophy, Department of Civil Engineering and Operations Research, Princeton University, 1996.
[3]     Tadepalli, T.P., Interactive computational tools for simulating linear dynamic response and nonlinear quasi-static damage in composite structures, thesis submitted for the degree of Master of Science, Department of Civil Engineering, University of Mississippi, 2003.
[4]     Bathe, K.J., *Finite Element Procedures*, Prentice-Hall: Upper Saddle River, NJ, 1996.
[5]     Lubliner, J., *Plasticity Theory*, Prentice-Hall, Inc., 1998.
[6]     Swann, C.T., Mullen, C.L., Hackett, R.M., Stewart, R.K. and Lutken, C.B., Evaluation of earthquake effects on selected structures and utilities at the University of Mississippi: A mitigation model for universities, final report submitted to MEMA, 1999.
[7]     Li, Q.M. and Meng, H., Pressure-impulse diagram for blast loads based on dimensional analysis and single-degree-of-freedom model. *ASCE Journal of Engineering Mechanics*, **128(1)**, pp. 87-92, 2002.
[8]     Design, materials, and connections for blast-loaded structures, ABS Consulting Ltd, Research Report 405, Health and Safety Executive, 2000.
[9]     Natural hazard mitigation plan of the University of Mississippi, Lafayette County, Mississippi, Center for Community Earthquake Preparedness, 2006 (www.olemiss.edu/orgs/ccep).

# Blast-resistant highway bridges: design and detailing guidelines

G. Williams[1], C. Holland[1], E. B. Williamson[1], O. Bayrak[1],
K. A. Marchand[2] & J. Ray[3]
[1]*Department of Civil, Architectural, and Environmental Engineering, University of Texas at Austin, USA*
[2]*Protection Engineering Consultants, USA*
[3]*US Army Corps of Engineers, Engineer Research and Development Center, USA*

## Abstract

The design of bridges to resist blast loads has become an international concern in recent years. Data from the US State Department indicate that violent attacks against transportation targets have increased worldwide over the last decade and that highway infrastructure has been the most frequently attacked transportation target. Since September 11th, 2001, increased emphasis on bridge security has raised awareness in the engineering community that bridges and other transportation structures should be designed to better respond to potential terrorist attacks. The fact that many bridges provide open access, carry thousands of motorists, and may have symbolic importance makes them attractive targets, and the success of recent terrorist bombings on bridges during the ongoing "war on terror" highlight the vulnerability of these structures. This paper presents preliminary results and observations from blast tests on concrete bridge columns conducted during a US national study to develop design and detail guidelines for blast-resistant highway bridges.
*Keywords: blast, bridge, column, concrete, explosive, terrorism.*

## 1   Introduction

Structural engineers have the responsibility of designing strong, durable structures that are able to resist extreme loading scenarios without collapsing. Past research on blast-resistant designs focused primarily on buildings, but the attention of the structural engineering community is now turning to highway bridges. Although ordinary highway bridges may not seem like probable terrorist targets, historical evidence suggests otherwise. A confiscated Al Qaeda training manual states that terrorist goals include "destroying and blasting bridges leading into and out of the city" in order to "strike terror [into the hearts] of the enemies" [1]. The success of recent attacks on overpass bridges in Iraq illustrates the realization of these goals, and historical data confirm that terrorists' desire to attack ordinary bridges spans many years. A report from the Mineta transportation institute [2] indicates that 53 terrorist attacks specifically targeted bridges between 1980 and 2006, and 58% of bridges targeted worldwide and 35% of bridges targeted in industrialized nations during that time were highway bridges other than signature bridges. Considering that 60% of all attacks on transportation targets during that time were bombings, a bombing of an ordinary highway bridge is a realistic scenario, and structural engineers need recommendations for blast-resistant bridge design. This paper presents preliminary observations from experimental research on blast-loaded bridge columns, and it outlines ongoing analytical work to develop design guidelines for blast-resistant bridges.

## 2 Test method

Highway bridges can vary significantly in size and configuration, and each has many structural members that contribute to the global response of the structure. Although many bridge types and structural components deserve attention, establishing the behavior of blast-loaded bridge columns will provide the greatest current contribution to the bridge community as a whole. Bridge columns are essential to nearly all bridges and bridge types, and they are arguably the most important structural element in a bridge. Many bents have only one column, which means failure of a single column could initiate collapse of an entire bridge. Elevated interstate highway interchanges with single column bents are especially vulnerable, and the collapse of the highest superstructure in one of these systems may mean failure of all those below. Additionally, bridge columns typically offer unrestricted access to the public, making them attractive targets for potential attackers. Thus, understanding their response to blast loads is essential, and this research investigates the influence of splice location, cross-section shape and size, and transverse reinforcement type and spacing on the behavior of blast-loaded bridge columns.

### 2.1 Specimens

Individual state departments of transportation (DOTs) govern bridge construction throughout the United States. As a result, design preferences can vary within the national bridge community, and the results of this research are intended to benefit a broad range of design practices. Although standards do vary throughout the nation, consultations with state DOT design guidelines and representatives show that general trends do exist, and the specimens in this study represent the most commonly used bridge column design parameters.

Table 1:    Design parameters for concrete bridge column specimens.

| Column Label | Shape | Diameter ft (m) | Longitudinal Reinforcement Ratio | Transverse Steel Type | Transverse Steel Design |
|---|---|---|---|---|---|
| 1A1 | Circular | 1.5 (0.46) | 1.04 | Hoops | Typical |
| 1A2 | Circular | 1.5 (0.46) | 1.04 | Hoops | Typical |
| 1B | Circular | 1.5 (0.46) | 1.04 | Spiral | Typical |
| 2A1 | Circular | 2.5 (0.76) | 1.13 | Hoops | Typical |
| 2A2 | Circular | 2.5 (0.76) | 1.13 | Hoops | Typical |
| 2B | Circular | 2.5 (0.76) | 1.13 | Spiral | Typical |
| 2-seismic | Circular | 2.5 (0.76) | 1.13 | Spiral | Seismic |
| 2-blast | Circular | 2.5 (0.76) | 1.13 | Spiral | Blast |
| 3A | Square | 2.5 (0.76) | 1.18 | Ties | Typical |
| 3-blast | Square | 2.5 (0.76) | 1.18 | Ties | Blast |

The testing program contains 10 half-scale specimens, and table 1 presents selected details of each specimen. The research plan emphasizes circular cross-sections because they are the most common cross-section used today in US transportation infrastructure. The half-scale specimens include three 1.5-ft (0.46-m) diameter round columns, five 2.5-ft (0.76-m) diameter round columns, and two square columns with edge widths of 2.5 ft (0.76 m). All specimens have a total height of 11.25 ft (3.43 m). Five circular columns and one square column employ typical DOT designs. The longitudinal reinforcement ratios of these specimens remain constant for all cross-section sizes and shapes, while the transverse reinforcement varies to study its influence on performance. All longitudinal reinforcement in the columns have splices near the base using conventional construction detailing, except columns 2B and 2-seismic, which have no rebar splices, to examine the effect of splice location on column response. The design of one column was based on seismic standards to investigate the influence of larger transverse reinforcement ratios, and two columns are special blast-resistant columns designed specifically for this study with significant increases in the transverse reinforcement volumetric ratio.

## 2.2 Boundary conditions and reaction structure

The boundary conditions selected for these blast tests are those of a propped cantilever, which are a fixed condition at the base of the column and a pinned condition at the top. These boundary conditions model a scenario in which the location of an explosion is directly beneath the deck and directly to the side of a column. Thus, the pinned condition at the top of the specimens models the superstructure, which is essentially axially stiff along its primary axis, and the fixed condition at the base of the specimens models the column foundation. The axial loads experienced by bridge columns in service typically do not exceed the balance point of the column, and any applied axial load only improves the response. Thus, testing specimens as propped-cantilevers with no applied axial load is conservative and logistically desirable.

A specially constructed reaction structure provides the desired boundary conditions. This reaction structure consists of a steel frame of 8-in × 8-in × 0.625-in (20.3-cm × 20.3-cm × 1.59-cm), A500 Grade B structural steel tubes cast in a 29-ft × 14-ft (8.84-m × 4.27-m) reinforced concrete slab. The steel frame provides the pinned connection at the top of the specimens, and the concrete slab provides the fixed boundary condition at the base of the specimens. Fig. 1 shows a schematic of the reaction structure and column, and Fig. 2 shows a picture of the reaction structure with a specimen prior to testing.

The slab is predominantly 2-ft (0.61-m) thick, but the section of slab nearest the explosion where the column foundations rest is 2.5-ft (0.76-m) deep. This front section has a 5-ft, 2.5-in (1.59-m) × 7-ft (2.13-m) cavity in which the columns rest. Each column sits on a 2.5-ft × 2.5-ft × 5.0-ft (0.76-m × 0.76-m × 1.52-m) foundation that fits into this cavity, and high-strength, quick-set grout fills the gaps in the front and the back of the cavity to provide the fixed restraint. The cavity also has room on each side of the column foundations for instrumentation wires to exit the foundations and connect with cables connected to the high-speed data acquisition system.

## 2.3 Blast tests

The construction of the specimens began in August 2007 and finished in September 2007, and testing commenced in early October 2007 and extended

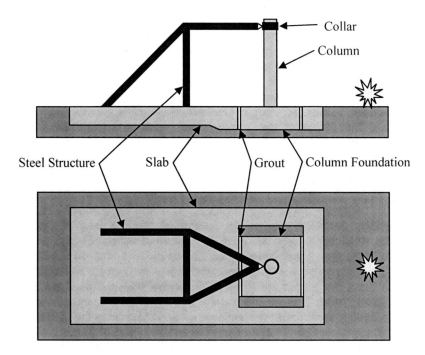

Figure 1:    Schematic of reaction structure and test setup.

Figure 2:     Picture of reaction structure and column before the test.

over a 16-day period. As previously mentioned, the column foundations lowered into the cavity in the front of the slab. Grout placed between the front and back faces of the column foundations and the reaction slab provided the fixed conditions at the base, and a steel clamp attached to the top of the columns that bolted to the steel reaction frame supplied the pinned restraint at the top of the columns. This method of connecting the column to the reaction structure provided the desired boundary conditions, while allowing for quick column removal and replacement.

Testing of the columns was divided into two different series. The intent of the first test series was to evaluate the overall performance of each column design. For this purpose, the charge weights and standoff distances were selected to avoid significant spall and breach damage, while identifying the dominant mode of response – whether shear, flexure, or a combination of both – for the set of design parameters used in each column. Afterwards, a second test series examined the spall and breach capacities of selected specimens. The columns selected for these additional tests had minor flexural or shear cracking only, and thus the damage existing prior to these tests did not greatly influence the results. In both test series, ANFO was used for the explosive charges; this paper does not disclose the exact charge weights and standoff distances for security purposes.

Three pressure gages recorded overpressures at 37 ft (11.3 m), 52 ft (15.9 m), and 76 ft (23.2 m) away from the charge, and each test yielded 6 channels of strain data for the reinforcing steel. The two specimens with no splices had three strain gages on the transverse steel and three strain gages on the back-face flexural steel (i.e., tension reinforcement for initial response), and the eight specimens with longitudinal reinforcement splices had two strain gages each above and below the splice on the back-face longitudinal steel (i.e., four total gages on the longitudinal reinforcement) and two strain gages on the transverse reinforcement near the base of the column. High-speed cameras visually captured the response, and after each test, the field team inspected the specimen, identified and recorded damage, marked and sketched crack patterns, and thoroughly photographed all observations.

# 3   Initial observations and results

The visual and recorded data show the dominant modes of response for a given set of column parameters, charge weight, and standoff distance. Several initial observations illustrate general trends in performance related to a column's section properties, and additional analysis will provide detailed recommendations for bridge column design. This paper cannot describe the performance of each column in great detail due to security concerns, and it only provides an overview of basic observations.

## 3.1 Shear and flexure tests

Prior to testing, shear at the column base (both direct and sectional) was anticipated to control the performance of the blast-loaded columns. Although the principle response modes varied depending on section properties, charge weight, and standoff distance, base shear did clearly dominate the response in most cases. Extensive shear cracking patterns were evident in most specimens, and these specimens showed little to no flexural cracking. A few tests with larger charge weights or smaller standoff distances resulted in extensive shear damage at the base of the column. Figure 3 shows extensive shear damage at the base of one of the specimens. As planned, these specimens experienced essentially no spall or breach damage during the initial shear and flexure tests.

The scaled standoff distances forced a few specimens to complete shear failure, but some columns appeared to retain load-carrying capacity. Although direct shear dominated the response in most cases, specimens with adequate shear capacity performed very well and exhibited clear cracking patterns indicative of combined flexural and shear behavior. Figure 4 shows flexural cracking on the back face (i.e., tension side for initial response) of a specimen.

Overall, the half-scale bridge columns performed more robustly than expected. Trial specimens tested to establish appropriate scaled standoffs had only superficial damage, and the blast intensities of subsequent tests had to be increased to obtain desired levels of damage. This observation shows that bridge columns may have greater capacity than previously thought; however, a few design parameters can change to improve response, two of which include increasing the volumetric transverse steel ratio and eliminating longitudinal reinforcement splices in vulnerable regions. Furthermore, cross-sections with continuous (i.e., spiral) shear reinforcement performed better than columns reinforced with the same percentage of steel using hoops.

Figure 3:     Extensive shear damage at the base of the specimen.

Figure 4:     Flexural cracking on the back face of the specimen.

## 3.2  Spall and breach tests

As previously mentioned, additional tests examined the spall and breach capacities of six columns.  The specimens used for these tests sustained only light damage during the first series of shear and flexure tests, and the minor cracking damage present before the second series of tests did not significantly affect the results.  Initial observations indicate that current methods available to predict spall and breach of concrete walls do not apply to columns.  Due to security restrictions, this paper cannot provide additional information about these tests.

## 4   Future work

The information obtained during these 16 blast tests provides a valuable foundation on which to build blast-resistant bridge columns, and analytical parameter studies will further contribute to the understanding of how selected design parameters influence column response.  To that end, the strain gage data and visual records from the experimental blast tests described in this paper will permit the calibration of 3D nonlinear, finite element analyses in LS-DYNA [3] that will study the influence of cross-section shape and size, longitudinal reinforcement bar size and ratio, and transverse steel bar size and spacing on bridge column response to blast loads. The combined results of the experimental tests and the analytical parameter study will help establish design recommendations for blast-resistant bridge columns and help calibrate single-degree-of-freedom analysis tools for design office use.

## Acknowledgements

The authors would like to thank the professors and graduate students of the University of Texas at Austin, United States for their instruction, time, and dedication to this research; Protection Engineering Consultants in Texas, United States for the successful design and management of the blast tests; the Southwest Research Institute in Texas, United States for their field work and expertise in explosive engineering; and the National Cooperative Highway Research Program (NCHRP) of the United States for the funds needed to conduct this research. The information contained in this paper reflects the opinions of the authors and not necessarily those of the sponsor.

## References

[1] Military Studies in the Jihad Against the Tyrants, Al Qaeda Terrorist Training Manual captured in Manchester, England; United States Department of Justice (USDJ) Online.

[2] http://www.usdoj.gov/

[3] Jenkins, B.M. & Gersten, L.N., *Protecting Public Surface Transportation against Terrorism & Serious Crime*, Mineta Transportation Institute: San Jose, California, 2001.

[4] Livermore Software Technology Corporation (LSTC), *LS-DYNA Keyword User's Manual, Version 971*, Livermore Software Technology Corporation: Livermore, California 2007.

# Approximation of blast loading and single degree-of-freedom modelling parameters for long span girders

J. C. Gannon[1], K. A. Marchand[2] & E. B. Williamson[3]
*[1]Walter P. Moore & Associates Inc., USA*
*[2]Protection Engineering Consultants Inc., USA*
*[3]The University of Texas at Austin, USA*

## Abstract

In this paper, the modelling of long-span girders under blast loads is presented. Specifically, spans in the range of 80–160 feet, on the order of those used for typical highway girder bridges, are considered. Topics addressed in this paper include (1) applicability of a uniform equivalent load to model blasts acting on long spans, (2) mathematical development of resistance functions and dynamic transformation factors for beams subjected to multiple distributed loads, and (3) comparisons of dynamic single degree-of-freedom analyses using both a work-equivalent uniform load and an approximation using three distributed loads of variable lengths relative to a detailed representation of the blast load profile as a function of position and time using finite element analyses. Analytical studies showing the sensitivity of the results to variations in the assumptions used to determine the magnitude and length of the loading pattern are provided. Based on these studies, a new method for approximating the response of long-span girders subjected to blasts with small scaled standoffs is proposed, which differs from the equivalent uniform load approach that is typically utilized. The new method is used to carry out parametric studies of bridge superstructure response predictions as part of research work performed for a state pool-funded bridge security project and an NCHRP project involving blast-resistant bridges.
*Keywords: blast load, uniform equivalent load, distributed load, bridge girder, single degree-of-freedom, SDOF, load-mass factor, bridge loading, long span.*

# 1　Introduction

Determining the response of a structure or structural component to blast loading can be a challenging task due to the fact that blast loads vary with both time and position and structures respond dynamically, often with large deformations, in response to these loads. A wide range of analytical techniques, ranging from detailed, coupled multiple degree-of-freedom (MDOF) finite element models to simple single degree-of-freedom (SDOF) models, can be used to provide information about structural behaviour. Depending on the resources available and the required fidelity of the results, a decision must be made about the most suitable analysis technique. Through the use of a sound set of assumptions, SDOF models can be effectively utilized to capture important characteristics of structural response while using a minimal amount of computational resources and analyst time. For these reasons, SDOF modelling is commonly considered the state-of-practice for modelling responses of simple components subjected to blast loading.

This paper addresses a refinement of assumptions made about characteristics of loading for bridge girders modelled as SDOF systems. Refinement is necessary to improve the quality of the results obtained by these simple models. The methods presented and discussed herein are intended to be consistent with the level of complexity typically employed with SDOF models. It is recognized, however, that some increase in analysis setup time over simple uniform loadings may occur.

When analyzing a structural component subjected to blast loads, model parameters that characterize the system, including the applied dynamic loading, must be specified. A SDOF model attempts to approximate the distributed mass and stiffness of a system or component through the use of discrete properties that account for key response characteristics, such as the maximum deflection at a critical location (e.g., midspan of a girder under transverse uniform loading). Previous work by Biggs [2] and others provides detailed information on the modelling of beams and slabs using an SDOF representation. In order to equate the actual system to the SDOF system, certain work equivalency factors must be determined and applied to the SDOF mass and load. These load and mass factors are based on the distribution of the actual mass and actual load on the real structure relative to the simplified SDOF model. This paper discusses calculation of these parameters for loading that is more complex than the uniform distributed loading that is typically considered in blast-resistant design.

# 2　Applicability of uniform equivalent loading for long span girders

For simplicity, a blast load acting over the span of a component is typically considered to act as a uniform load. The load magnitude and its variation with respect to time may be determined by any number of methods such as work equivalency, a weighted average over a subjected area, or simply selecting the largest pressure and impulse acting on the component. In actuality, depending on the geometry of the component and its surroundings, as well as the standoff and orientation of the blast source, blast loads are likely to vary over the area of the component being analyzed. While in some cases it may be quite reasonable to approximate an applied blast pressure as acting uniformly over a component, this assumption becomes unrealistic as the variation in loading becomes large.

The pressure acting on a surface as a result of a blast is related to the standoff from the blast and the angle of incidence of a line from the blast source to a point on that surface. For example, consider a long-span girder, on the order of 80–160 feet, such as those found in highway bridges, subjected to a blast located some at some distance perpendicular to its longitudinal axis. The standoff varies significantly from the midspan to a point at the end of the girder, and the angle of incidence also substantially changes. Both of these factors contribute to creating a significant pressure gradient over the length of the girder. With a relatively long structural member subjected to a significant pressure gradient, it is unrealistic to utilize a uniform load to approximate the actual behaviour. Analytical evidence of this concept is presented later in Section 4. The concept of using a single uniform load to approximate the distribution of pressure over a blast loaded girder is illustrated below in fig. 1.

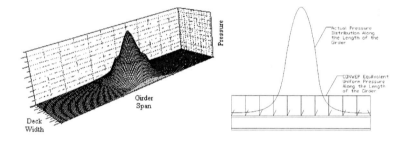

Figure 1:     Distribution of blast pressure over a long span girder.

## 3  Development of dynamic analysis parameters for SDOF beams subjected to multiple uniform loads

In order to perform an SDOF analysis of a girder, system parameters such as stiffness, mass, ultimate resistance (load causing formation of a collapse mechanism), and equivalency factors equating the real system to the idealized system must be determined. The values of these parameters are subject to assumptions made about the displaced shape of the component under loading. Several choices exist for formulating these parameters. This research uses the displaced shape of the component under a static load of the same form as the dynamically applied blast load. The selection of this displaced shape corresponds to the recommendations made by Biggs [2] and is commonly accepted as the current state-of-practice. Examples of parameter calculations, along with tables of various system properties and transformation factors, can be readily found in several sources such as dynamics textbooks (e.g., Biggs [2]) and the Army TM5-1300 [4], a valuable reference for analysis of structures subjected to blast loading.

This research focuses on formulation of system parameters for beams subjected to a series of uniform loads of different magnitudes. Development of these parameters is performed in a manner consistent with the methods employed by Biggs [2] for one-way components. The first step in the process requires calculation of the static displaced shape. Fig. 2, shown below, is an illustration of the loading condition which is used to replace the uniform equivalent load. Lines are used to differentiate the regions associated with the different load magnitudes. Different continuous functions within each load region describe the variation in transverse displacement with position. Continuity of the beam can be used to relate the expressions in each of the different segments.

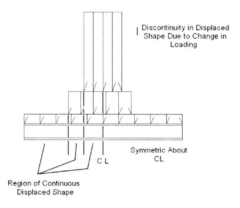

Figure 2:     Loading condition diagram.

The SDOF stiffness is calculated as the reciprocal of the peak deflection of the static displaced shape. Multiple stiffness values are determined based on changing displaced shapes that occur as a result of the formation of plastic hinges. Resistance levels that form the bounds of these stiffness regions are calculated using plastic analysis techniques to determine the load level at which plastic hinges form. Depending on the selected boundary conditions (e.g., fixed or simple), two or three stiffness regions may exist. In order to effectively utilize these defining characteristics, load and mass transformation factors equating the real and idealized systems must be formulated. The factors account for the variation of mass and load over the displaced shape and are determined in accordance with eqns (1)–(3) below. In these equations, $M_e$ is the mass transformation factor, $m$ is the mass per unit length of the beam, $\phi(x)$ is the normalized displaced shape of the beam (i.e., peak deflection is scaled to a unit value), $L_f$ is the load transformation factor, $w$, $c$, and $p$ are the applied distributed loadings, respectively, in each region, $L$ is the beam length, and $L_{mf}$ is the load-mass transformation factor.

$$M_e = \int m \, \varphi(x)^2$$

(1)

$$L_f = \frac{\displaystyle\int_0^{nL} w \cdot \varphi(x) \cdot dx + \int_0^{(j-n)L} c \cdot \varphi(x) \cdot dx + \int_0^{(1-j-h)L/2} p \cdot \varphi(x) \cdot dx}{(w+c+p) \cdot L}$$

(2)

$$L_{mf} = M_e \, / \, L_f$$

(3)

Each integration involving the displaced shape $\phi(x)$ must be performed piecewise and correlated with the appropriate length over which that displaced shape is valid. The appropriate load acting over that displaced shape must also be used, and the resultant of the entire load is required in the denominator of Equation 2. The evaluated mathematical expressions, including the displaced shapes, are not shown here because of their algebraic complexity; however, they have been derived in a manner that allows for changing of the lengths over which the different uniform loads act in order to account for blasts of differing distributions. Development was performed separately for fixed and simple boundary conditions. Details of the resulting expressions for the various analyses parameters can be found in Gannon [5].

## 4    Comparisons of loading methods using SDOF and finite element modelling

Current practice for component analysis using SDOF systems is to utilize a uniform equivalent load as computed using a program such as CONWEP [3]. Comparisons with SDOF models that allow for a variation in the applied loads as defined above are presented here to illustrate the effectiveness of this technique. Additional comparisons are made to finite element beam models that allow for a detailed variation in the applied load to be prescribed. The program SBEDS is used to calculate the response of SDOF models of steel bridge girders subjected to the uniform and multiple uniform loadings.

To illustrate the differences in loading techniques for long spans, a steel girder with the properties shown in table 1 was subjected to TNT equivalent explosives of a magnitude on the order of a vehicle bomb. Standoffs of 12 and 20 feet were examined. The girders studied were assumed to be fully braced, and effects of the deck, such as added mass or composite action, were not considered.

The result of primary interest, which can be obtained from an SDOF analysis, is the midspan displacement history. In this case, flexural behaviour was modelled, and corresponding displacements were determined. Figure 3 is a plot of midspan displacement histories of an 80-foot girder subjected to 2000 pounds of TNT at a standoff of 20 feet determined using an SDOF model with a CONWEP [3] uniform equivalent load, an SDOF model using the multiple distributed loads presented previously, and an ANSYS-LSDYNA [1] beam model using a series of uniform loads taken as an average of the CONWEP [3] pressure and impulse distribution along the girder length (a different series than used for the SDOF model). An 8-foot width was used for the tributary width, and a 32-foot width was used to determine the CONWEP [3] reflecting surface area. A steel yield strength of 50 ksi under static loading rates was assumed. This value was modified for material over-strength and increases due to strain rate effects to give an effective steel yield strength of 62.5 ksi.

Table 1:     Properties of studied steel girder.

| Property | Value |
|---|---|
| Girder Depth (in) | 72 |
| Web Thickness (in) | 1 |
| Area (in$^2$) | 168 |
| Moment of Inertia (in$^4$) | 150408 |
| Plastic Section Modulus (in$^3$) | 4680 |
| Weight (lbs/ft) | 571.7 |

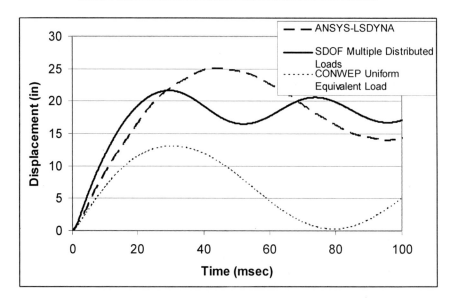

Figure 3:     Midspan flexural displacement history of an 80 foot girder subjected to different loading types.

Several girders were examined to provide a range of data points for comparison of the available loading techniques.  Girder parameters and explosive loadings were held constant, with standoffs of 12 and 20 feet considered.  In order to include the importance of the load variation with position, multiple span lengths were included in the study. Lengths of 80, 120, and 160 feet were studied. These lengths encompass the majority of spans of typical steel girder highway bridges for which the multiple distributed loading may be most appropriate. The CONWEP [3] reflecting surface was assumed to be the same length as the span under consideration. Table 2 shows the calculated peak displacements using the various loading alternatives under consideration. The values in this table include the peak displacements from each curve shown in fig. 3 above.

Table 2: Comparison of peak flexural displacement values of girders using various modelling techniques and loadings.

| Loading/analysis technique | Span | Standoff | Peak displacement | Error relative to ANSYS-LSDYNA |
|---|---|---|---|---|
| | (Feet) | (Feet) | (Inches) | (%) |
| CONWEP UEL[1] | 80 | 12 | 19.7 | -62.9 |
| Multiple Distributed Loading | 80 | 12 | 32.1 | -39.6 |
| ANSYS-LSDYNA | 80 | 12 | 53.1 | N/A |
| CONWEP UEL | 80 | 20 | 13.1 | -47.8 |
| Multiple Distributed Loading | 80 | 20 | 21.6 | -13.9 |
| ANSYS-LSDYNA | 80 | 20 | 25.1 | N/A |
| CONWEP UEL | 120 | 12 | 27.5 | -69.1 |
| Multiple Distributed Loading | 120 | 12 | 101.9 | 14.4 |
| ANSYS-LSDYNA | 120 | 12 | 89.1 | N/A |
| CONWEP UEL | 120 | 20 | 20.6 | -58.0 |
| Multiple Distributed Loading | 120 | 20 | 38.5 | -21.5 |
| ANSYS-LSDYNA | 120 | 20 | 49.1 | N/A |
| CONWEP UEL | 160 | 12 | 36.4 | -64.3 |
| Multiple Distributed Loading | 160 | 12 | 110.3 | 8.2 |
| ANSYS-LSDYNA | 160 | 12 | 102.0 | N/A |
| CONWEP UEL | 160 | 20 | 29.3 | -62.2 |
| Multiple Distributed Loading | 160 | 20 | 95.1 | 22.8 |
| ANSYS-LSDYNA | 160 | 20 | 77.4 | N/A |

[1] UEL denotes uniform equivalent load.

Review of the error relative to the peak midspan displacement determined by an ANSYS-LSDYNA [1] FEM, shown above in the last column of table 2, clearly demonstrates that it is inaccurate and inappropriate to use a CONWEP [3] equivalent loading to characterize a blast pressure distribution over a long span. The large pressure and impulse gradients are not effectively captured using the work equivalent method included within the software. It should be noted that it may be possible to more closely match the characteristic displacement using another form of uniform loading; however, further study would be required to devise a proper method for calculating the equivalent pressure and impulse.

The results shown in table 2 clearly indicate that the response predictions based on the multiple uniform loadings are more accurate than the corresponding values obtained using the CONWEP [3] uniform equivalent loading when compared to the ANSYS-DYNA [1] solutions. While a certain degree of error still exists, response predictions compare more favourably to the finite element analyses, and the level of effort required to achieve these results is comparable to that used when approximating the load as acting uniformly over the span.

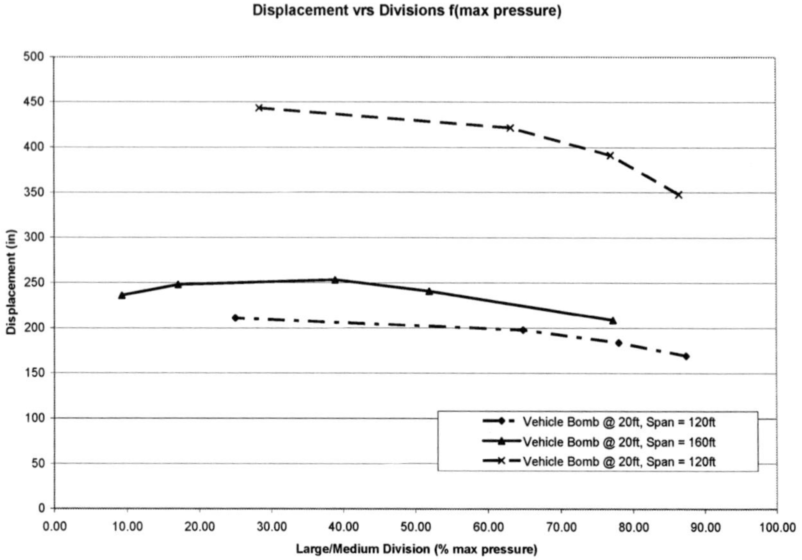

**Displacement vrs Divisions f(max pressure)**

Figure 4: Peak midspan displacement versus relative length of multiple distributed loadings.

One consideration which is of importance to the use of multiple distributed loadings is the points at which the pressure gradient is approximated by breaks from one uniform load to the next. The model presented in this paper uses three distributed loads and requires divisions at 2 locations (on each side of girder midspan because of the symmetry assumed in these examples). The average pressure and impulse must then be determined for each uniform load over the area in question. If the location of these breaks is modified, slightly different dynamic system parameters would be calculated, and, therefore, a different midspan displacement would be obtained. Fig. 4 illustrates different midspan displacements which would be calculated based on alternate selections of break points. In Fig. 4, the location of the break between the smallest and next largest distributed loads is held constant, and the division between the largest and next largest distributed load is represented on the x-axis as a percentage of the peak blast pressure. A chart such as the one shown below can easily be generated for a range of system configurations under study, and a break point can be chosen such that conservative answers are determined by SDOF modelling. This paper has used a break point between the largest and next largest magnitude distributed loads of 45% of the peak pressure. An alternate break value could be chosen, but the relative accuracy of the multiple distributed loading method, compared to ANSYS-LSDYNA [1] analyses, would be approximately the same despite slightly different displacements values for each system.

## 5  Conclusions

It has been shown that for the analysis of long-span girders, a series of uniform loads of different magnitudes is more appropriate for characterizing an applied blast load than a uniform equivalent load. Although mathematically the expressions for determining relevant system parameters are more cumbersome for a system subjected to multiple uniform loads, they can be easily written into spreadsheets or mathematical codes for quick evaluation. The added accuracy over a single uniform load, as demonstrated by comparison to FEM using a more detailed description of load, may be valuable in analytic studies or design where confidence in results can be used to reduce conservatism. The method of load description presented in this paper was effectively utilized in parametric studies of bridge girders under blast loading for a recent pool-funded study conducted by The University of Texas at Austin under the supervision of Dr. Eric Williamson. The purpose of that study was to determine effective methods of mitigating risk of failure of in-service highway bridges and to identify effective design concepts which could be used to improve blast resistance of future structures. A large number of SDOF models were evaluated to form a basis for the evaluation of different design and retrofit concepts. This method of blast pressure description was useful in part because it provided a method of load relief by using a load path approach. Because the analysis considered load variation over the length of a girder, failure of portions of the deck, which were loading the supporting girders, was readily modelled. A more accurate representation of the blast load acting on a girder or set of girders allows for an improved understanding of structural response and a more useful set of analytical results. Furthermore, because SDOF models are utilized, the time needed for analysis is less than that which would be needed for detailed finite element studies. Thus, the proposed blast load modelling alternative offers the advantage of increased accuracy over typical SDOF analyses while maintaining simplicity in the model development. While more advanced methods of analyses are needed for more accurate response predictions, the proposed method is well suited for initial design and for parametric studies that are often essential to the design process.

## References

[1]  ANSYS LSDYNA Version 10.0, ANSYS Inc., 2005.
[2]  Biggs, J.M., *Introduction to Structural Dynamics*, McGraw-Hill Book Company: New York, NY, 1964.
[3]  CONWEP Version 2.1.01, United States Army Corps of Engineers, (USACE) U.S. Engineering Research and Development Center, Vicksburg, MS, 2003.
[4]  Department of the Army, *Structures to Resist the Effects of Accidental Explosions*, Army TM 5-1300, U.S. Government Printing office, Washington, D.C., 1990.
[5]  Gannon, J.C., *Design of Bridges for Security against Terrorist Attacks*, Austin, TX, 2004.

# Simulation-based design of vehicles exposed to blast threats for improved occupant survivability

R. T. Bocchieri, S. W. Kirkpatrick & B. Peterson
*Applied Research Associates, Inc., USA*

## Abstract

Designing vehicles to protect occupants from explosive threats is complicated by the complex set of physics that occurs from the point of detonation to the response of the occupants. These physics include detonation chemistry, shock physics, solid mechanics, structural dynamics, nonlinear material behaviour, and human physiology and injury mechanics, among others. Consequently, vehicles are typically developed through an iterative process of destructive field testing in order to determine the level of blast protection. This testing is both time consuming and costly and does not ensure an optimized design. Recent advances in computational power and high-fidelity multi-physics computational tools now offer the alternative of performing Simulation-Based Design (SBD), similar to what is currently done for crash protection. Here, the explosive threat, its coupling to a vehicle, the vehicle structure, occupants or surrogates, and their coupling to the vehicle are all modelled in a single analysis. In this paper we describe an approach that applies state-of-the-art nonlinear finite element analysis to rapidly determine the survivability of vehicle designs and occupant injury potential when subjected to a buried charge or an above-ground improvised explosive device (IED). This integrated analysis capability allows iterative assessments to be performed as part of the design process. Modelling approaches for buried charges and IEDs and the coupling of the resulting loads on the vehicle are discussed. Finally, detailed sub-modelling of critical components such as occupant seating systems is demonstrated.
*Keywords: armoured vehicle, blast, mine, IED, occupant survivability.*

# 1  Introduction

Mine blasts and improvised explosive devices (IEDs) pose a serious threat to occupants of armoured vehicles. These threats result in a variety of insults to occupants: fire, fragment penetration, primary blast loading, and both global and local blast-induced vehicle motion. Vehicles must be able to reasonably withstand these threats in order to maintain crew survivability. Of these threats, designing vehicles to be more survivable against blast-induced global vehicle kinematic motion and high-rate local deformation poses a significant challenge. Improved armour solutions can reduce injuries from projectiles, for example, however they can also result in greater injuries from blunt-trauma related vehicle motions. These types of design tradeoffs interact at a vehicle system level.

During vehicle development, contractors typically provide prototype vehicles for destructive field testing in order to determine the level of blast protection. This testing is both time consuming and costly and does not ensure an optimized design. A variety of field conditions and other related factors tend to make each blast test unique to some degree. Researchers struggle with comparative testing of candidate blast protection articles and are usually left to rely on field experience and intuition when developing and finalizing protective designs.

Developing a virtual environment to evaluate competing protective designs is crucial. This environment would allow the designer to analyze and evaluate a large array of test conditions and prototypes. It would also offer a first level optimization of protective blast protection designs prior to committing to destructive field testing. This design methodology would significantly reduce development costs and produce a better vehicle.

# 2  Simulation-based design

The concept of a Simulation-Based Design (SBD) system is to apply numerical simulation methods that allow for the rapid iterative evaluation of various concepts in the design process. Such a system would allow the user to manipulate features of the design and get rapid feedback on the effects of changing these features. Features that a designer may wish to manipulate include vehicle geometry, material properties, material thicknesses, and threat types and locations. The feedback to the user may include various evaluation criteria such as damage to the vehicle or injury to human occupants. Ideally, a SBD system would offer a convenient interface, run quickly, and easily incorporate new vehicles, technologies, designs, or threats.

A simplified flow chart illustrating the main components of an SBD system is shown in fig. 1. Good engineering software tools already exist for many portions of the system for many applications. In many cases, development is needed to link them together in an automated fashion. A Graphical User Interface (GUI) would ideally be used to provide easy-to-use input menus. A computer-aided design (CAD) system is used to create the geometry of the entities to be modelled and automeshing software then converts the CAD description of the geometry into a computational mesh. At the core of this methodology is a physics-based solver (e.g., finite element (FE) code, Computational Fluid Dynamics (CFD) code) suitable for modelling the physical behaviour. Pre-determined performance measures are then extracted from the physical simulation and an optimizer selects the parameters for the next design iteration. The degree to which this process can be automated will depend on the complexity of the system to be designed.

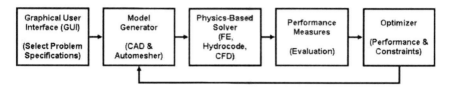

Figure 1:    Simulation-based design flow chart.

Numerical modelling of the effects of blasts on vehicles involves modelling the blast source, coupling the loading from this source to the vehicle structure, and simulating the vehicle response and that of the occupants. Modelling methods range from empirical models of the mine blast and analytical models for simple vehicle structures to fully coupled simulations using multi-physics computational codes. Because of the demonstrated accuracy of these codes and the relatively cheap computational power in recent years, use of multi-physics codes can now be used as the primary solver for performing simulation-based design of armoured vehicles.

The vehicle blast demonstrations in this paper use LS-DYNA as the primary solver. LS-DYNA is a commercially available nonlinear explicit finite element code for the dynamic analysis of structures [1]. LS-DYNA is developed and supported by the Livermore Software Technology Corporation (LSTC) and is used for a wide variety of crash, blast, and impact applications. LS-DYNA has several unique capabilities for application to blast simulation of armoured vehicles. These included numerical techniques that enable modelling of the blast threats: Smooth Particle Hydrodynamics (SPH) and Arbitrary Lagrangian-Eulerian (ALE).

## 3   Modelling of threats

Modelling of explosive threats is complex and high-fidelity simulations require a code that includes multi-physics capabilities. Important mechanisms to model for these threats include detonation physics and shock wave propagation, soil and or metallic casing mechanics and subsequent expansion of the detonation products and soil/fragment ejecta. Following are examples of modelling methods used for shallow-buried explosives and a representative IED.

## 3.1 Shallow-buried explosives (mine blast)

In order to develop and validate a generic mine threat model, an experiment performed by Defence R&D Canada (DRDC) with an offset aluminium armour plate was selected [2]. In the experiment, a 6.3-kg bare explosive charge was buried in 5 cm of soil. A 3.175-cm thick aluminium (Al 5083) plate was placed on a support stand 40.64 cm from the top of the soil. A steel support and 10 metric tons of mass were placed on top of the aluminium plate to simulate proper vehicle weight. The LS-DYNA model developed for this test configuration is shown in fig. 2. The explosive charge and surrounding dry sand were modelled using SPH elements. The far field sand is modelled using Lagrangian elements.

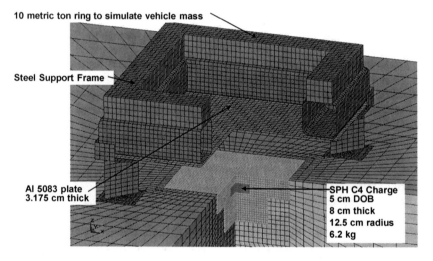

Figure 2:    LS-DYNA geometry for the DRDC experiment.

The computed plate deformation is shown in  fig. 3, and matches well with the published experimental results. The experiment showed a 30 cm displacement at the centre of the plate compared with 35 cm for the calculation. Both the experiment and the calculation showed a 5 cm edge displacement. Overall, the computed shape of the plate matches experimental observations.

## 3.2 Improvised explosive device (IED)

Cased explosive fragmenting weapons, such as an artillery shell, are commonly used in IEDs. Modelling such a weapon is complicated by the effects of the fragmenting case. The detonation of the explosive fill is initially confined in the case, and fragments the case as it expands. This fragmentation process develops during the detonation and initial expansion of the explosive products. Case material is rapidly expanded outward as the fragmentation progresses. For objects very close to the weapon, they are impacted by more intact components of the casing. The load on these objects more closely resembles the impact of a continuous material, rather than discrete fragments. At far distances, the fragmentation process has completed and the fragments have had time to spread out. Loading on far-field objects is best described as that from discrete fragment impacts.

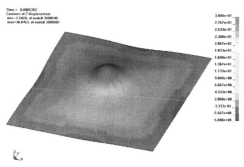

Figure 3: Plate deformation from LS-DYNA SPH calculation.

Loading from the IED comes from significant contributions of both the airblast and fragment impacts. The pressure-time history and shape of this airblast wave front is dependent on the explosive weight, case geometry and material properties, and standoff distance. A simplified approach to modelling the effect of the case-fragmenting IED is shown in fig. 4. In this model, a Lagrangian mesh was used for the steel case and an ALE mesh was used for the high explosive (HE) fill inside the case as well as the surrounding air. An alternative modelling approach would be to use an ALE mesh for the case as well. For close-in applications, the ALE approach is appropriate as case fragmentation has not occurred.

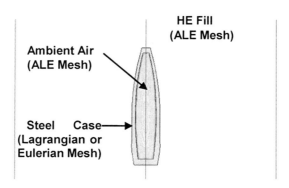

Figure 4: Model configuration for the IED detonation simulation.

An example simulation was performed of the IED interacting with a plate of Rolled Homogenous Armour (RHA), a material commonly used in armoured vehicles. Fig. 5(a) shows the model geometry, where the steel plate is modelled with Lagrangian shell elements and the air, explosive, and steel case are modelled with ALE elements. The resulting damage to a one-inch and a one-half-inch plate from detonation of the IED is shown in figs. 5(b) and 5(c), respectively. Such simulations are useful for validating threat models of IEDs using results from relatively inexpensive experiments.

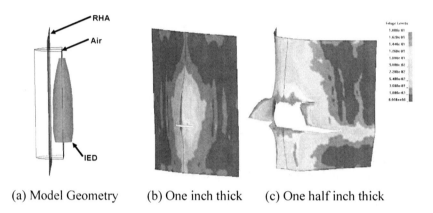

    (a) Model Geometry    (b) One inch thick    (c) One half inch thick

Figure 5:    RHA plate deformations from near-field IED simulations. (Contours of effective plastic strain are shown.)

## 4 Demonstrations of fully-coupled blast analyses of vehicles

For many applications, coupling of loads to a structure can be accomplished using the traditional 'uncoupled' or 'loosely' coupled approach. With this approach, loading is applied from separate empirical model or multi-physics analysis of an explosive threat. Such an approach is appropriate when the structural response time is greater than the blast load duration. A typical application of this method is for airblast loads on most engineered building structures. However, this assumption may not hold when lighter or thinner structures such as window systems or vehicle components are considered and impacted by a mixed phase flow (such as air/soil). As the characteristic response time of the structures or components of interest approach the characteristic loading duration, the uncoupled approach can produce misleading results (e.g., [3]).

A fully-coupled analysis provides simultaneous solution of fluids, soil, casing and structural response for as much as every time step of a computational analysis. Such an analysis is needed when the structural response time is comparable or less than the blast load duration. A significant advantage of a fully-coupled analysis regardless of response time is that it provides the capability to perform a single end-to-end simulation of the event in one analysis.

Mine blast has been shown to be one application where this approach can be needed for some vehicle structures. Two examples of fully-coupled analyses from the threat models discussed are summarized here.

## 4.1 Mine blast beneath an armoured personnel carrier

The first example of a fully-coupled analysis appropriate for SBD is that of an armoured personnel carrier with a land mine beneath a track. Positioning of the M113A3 on the explosive charge and ground model is shown in fig. 5. The vehicle is aligned so that the charge under the second road wheel on the left-hand side of the vehicle. The charge zone, including the local area soil and charge, are modelled using SPH elements, which was previously described. Beyond the charge zone, the global soil area was modelled using Lagrangian elements.

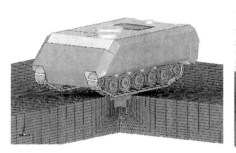 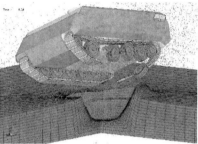

Figure 6:    LS-DYNA SPH calculation of a buried bare charge against an M113A3.

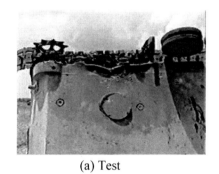 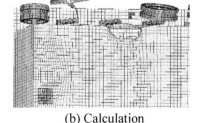

        (a) Test                   (b) Calculation

Figure 7:    Comparison of tested and calculated damage to M113A3.

A comparison of the calculated damage to the hull to a corresponding buried charge test is shown in fig. 6. The calculated damage matches the experiment quite well. In both the experiment and calculation the first two road wheels have failed, and there is a large deformation of the hull near the second road wheel.

These calculations give an example of how a SBD system can be used to quickly analyze and compare different threats against a vehicle.

## 4.2 Near-field IED simulation on an 'up-armoured' Chevrolet C2500 pickup truck

The second example of a fully-coupled analysis appropriate for SBD uses the IED threat model developed against a notional up-armoured pickup truck for a near-field detonation. The vehicle modelled in this simulation was the up-armoured 1989 Chevrolet C2500 pickup truck. The pickup model was originally developed by the U.S. Federal Highway Administration (FHWA) for highway crashworthiness studies [4, 5]. It has been notionally 'up-armoured' here to provide a more realistic representation of an armoured vehicle response to an IED or mine blast.

In order to reinforce the cab, the components shown in grey have been replaced with 0.4 inch RHA. A 1.0 inch thick plate of RHA, shown in tan, has been added under the cab in order to protect against mines and IEDs. To support the increased weight of this armour, the thicknesses of frame components, shown in red, have been doubled. The new total weight of the vehicle is 3.08 tons (a 62% increase from the un-armoured weight). Naturally, a complete model would require modifications to the suspension and possibly other components to support the added weight. Recall that the bottom armour plate thickness offers significant protection against this IED even at zero standoff.

A model of the IED was placed directly beneath the crew cab of the pickup and centred between the vehicle doors, as shown in fig. 8. The ALE mesh encompasses much of the vehicle cab. In this simulation, the IED explosive products and case material have been represented in the ALE mesh, while the pickup is comprised of Lagrangian shell elements. The IED has been placed on top of a rigid ground, best representing a surface such as concrete.

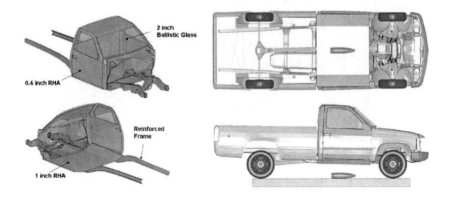

Figure 8:    IED location beneath the up-armoured C2500 pickup truck.

The vehicle response to the IED detonation is shown in fig. 9 with contours of pressure superimposed on the model. The vehicle is viewed in the figure with a cutaway through the centreline of the pickup. In the simulation, the front of case material is completely stopped by the 1 inch plate of RHA. However, a pressure shock does propagate through the armour plate, through the cab floor, and into the occupant compartment. Associated damage to the vehicle is also shown in the figure with contours of plastic strain. There is significant plastic deformation in both the armour plate and the vehicle frame.

## 5   Occupant injury assessment

To ensure the survivability of occupants in armoured vehicles requires many forms of protection. Typically, armour is applied to provide protection against ballistic and blast threats. However, as the vehicles become lighter the same blast threats impart larger gross vehicle motions. An armour solution may be designed at the material-level to stop the ballistic threats, but the vehicle is still thrown so violently that the occupants are still injured or killed from the blunt-trauma injuries induced by the severe acceleration environment. Because the response of an entire vehicle affects the accelerations of the occupants, including occupant models to assess injury potential in a SBD program is critical. Following is an example where computational models of Hybrid III dummies are used in a full vehicle analysis.

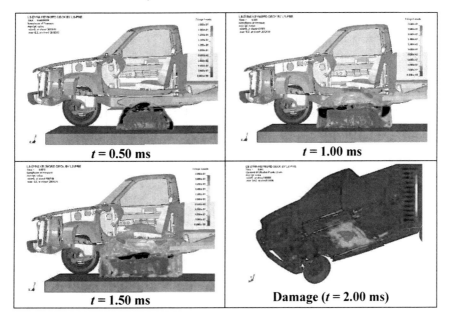

Figure 9:    Pressure contours and damage from an IED detonation beneath the C2500 pickup truck.

Simulation of the M113A3 exposed to a mine blast discussed in the previous section included explicit modelling of six occupants placed in the passenger compartment of the vehicle [6]. Fig. 10 shows the vehicle and occupant response at various times. At early times, the loading is transmitted through the floor to the lower extremities of the occupants. The subsequent loading results in uncontrolled occupant motions that are upward and out of their seated positions. From this occupant response, several injury criteria can be evaluated.

In this example, all of the occupants experienced a loading below the threshold Head Injury Criterion (HIC) value of 1000 (15% risk of an Abbreviated Injury Scale (AIS) 4+). The occupant closest to the mine (forward left position) had a HIC level of approximately 980. The head injury potential dropped rapidly for the other occupants. Subsequent motion of the unrestrained occupants also caused head strike injuries. The left middle occupant, for example resulted in a HIC level of approximately 700.

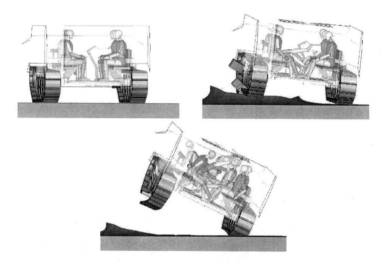

Figure 10:    M113 mine blast simulation with occupants.

## 6    SBD of seating systems using sub-modelling

The SBD approach discussed thus far has been on the assessment of the entire vehicle in response to given threats. Full vehicle models were constructed with sufficient fidelity to assess the vehicle damage while maintaining reasonable run times. As a result, the models are not necessarily suitable for detailed aspects of the vehicle structure, such as detailed design of a shock-absorbing occupant seat system, or modelling of localized threats (e.g., penetration or an explosively formed projectile (EFP). Sub-model development allows higher fidelity simulations aspects without affecting the run times of the global vehicle model.

The use of sub-modelling during SBD is demonstrated here for a vehicle seating system. Seats frequently contain detailed components, such as energy-absorbing features, that are critical to the seat performance. Evaluations of these components required detailed modelling procedures that cannot be efficiently performed inside of global vehicle response simulations. In addition, the design process for the seating system is commonly performed independently from much of the vehicle design effort. Sub-modelling is a good approach to address these issues. This modelling approach is valid to assess the response of any critical equipment mounted to the vehicle interior.

In the first stage of the vehicle sub-modelling analysis, the seat design may not be know. But, it is important to include the occupants and seat in global simulations as they affect the vehicle response to the threat. For this purpose, an estimated low-fidelity model of a seating system is developed to incorporate in a full vehicle model in order to perform global simulations. Once the global simulations and vehicle design are complete, detailed design and modelling of the seating systems are then performed.

An example sub-modelling approach for the M113 and a prototype seat system is shown in fig. 11. In the global simulation, the response of the occupant crew compartment is extracted and saved. A simplified sub-model of the crew compartment is then constructed with detailed models of potential seat designs and the compartment response from the global analysis is applied. Modelling the entire compartment is important because the occupant response may involve interaction with surrounding vehicle structures and not just the seat structure.

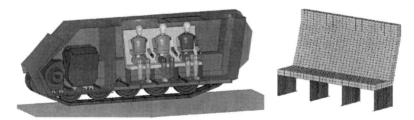

(a) Cutaway of M113 showing occupant compartment with simplified seat.

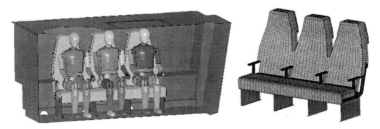

(b) Cutaway of the occupant compartment for sub-modelling with a detailed seat.

Figure 11:    Cutaway views of global and sub-models of the M113.

With the faster run times of the sub-model, many design iterations are feasible. The seat can be modified and simulations performed with the new design. Once the final design has been determined the global model with the detailed component could be run to confirm the results.

## 7 Conclusion

A well-run SBD vehicle development program utilizes various types of software to automate the development of computational models, simulate response of the vehicle, extract performance metrics and optimize the design. The degree to which this process is automated depends on the complexity of the system to be designed. At the core of this methodology is a physics-based solver suitable for modelling the physical behaviour and the development of accurate threat, vehicle, and occupant models.

In order for the SBD approach to be successful, representative threats need to be defined carefully. Then models for these threats need to be developed and validated. Validation can be performed using simple experiments such as blast tests against flat plates or simple structures and with devices to measure the loads and load distributions from the threats. Vehicle models need to be developed with sufficient fidelity to simulate desired behaviours (armour failure, suspension behaviour, etc.). Component tests can be used to validate the model fidelity. Finally SBD trade studies can be conducted with threats at prescribed locations. Structural response metrics and occupant injury measures are extracted and compared against requirements. Once the initial investment of developing these models has been made, the cost savings of the SBD approach compared to destructive vehicle testing quickly becomes evident.

## Acknowledgement

Support for this research by the US Army RDECOM under the direction of Richard C. Goetz is gratefully acknowledged.

## References

[1] LS-DYNA Keyword User's Manual, Livermore Software Technology Corporation, Version 970, April 2003.
[2] Williams, K., Validation of a loading model for simulating blast mine effects on armoured vehicles. *7th Int. LS-DYNA Users Conf.*, 2002.
[3] Peterson B.D., Kirkpatrick, S.W. and Bocchieri, R.T. Advances in finite element simulation of blast response to structures", *48th AIAA/ASME/ASCE/AHS/ASC Structures, Structural Dynamics and Materials Conference*, Honolulu, Hawaii, 2007.
[4] Tiso, P., Plaxico, C., Ray, M. and Marzougui, D., *An Improved Truck Model for Roadside Safety Simulations: Part II – Suspension Modelling*, Transportation Research Record No. 1797, pp. 63-71, 2002.

[5]  National Crash Analysis Centre Finite Element Model Archive, http://www.ncac.gwu.edu/vml/models.html.

[6]  Kirkpatrick, S.W., MacNeill, R.A. and Bocchieri, R.T., Development of an LS-DYNA occupant model for use in crash analyses of roadside safety features, Transportation Review Board, No. TRB2003-0002450, *Proc. of the 2003 TRB 82nd annual meeting*, Washington D.C., 2003.

# Blast protection in military land vehicle programmes: approach, methodology and testing

M. Müller, U. Dierkes & J. Hampel
*IABG Lichtenau, Dept. VG23 – Land Systems, Germany*

## Abstract

Personnel safety is crucial in operations where mines pose a threat. In peacekeeping and peace-enforcing operations, occupant protection is given top priority. During the past decade up to now, the IABG Defence and Security Department has been supporting the German MoD and Industry in most national and international vehicular mine protection and improvised explosive device (IED) protection programmes with independent engineering consultancy services and basic research. The original work shifted from the focus on protection against blast mines to a combined protection against blast and projectiles. With regards to the field of terrorist (IED) attack protection, the insights in blast mine protection still remain the basic background at which protection measures can be illustrated. This paper gives a brief overview about the blast mine protection field. Starting with general structural effects caused by a blast detonation, special focus is placed on the occupant loading and how it can be reduced. A brief description of simulation methods and injury criteria which are in focus for vertical loading conditions is therefore given. Beside vehicular qualification trials basic research on occupant protection by means of test rig configurations is presented. The test rigs for occupant safety systems TROSS I and TROSS II® give the opportunity to investigate different loading conditions and interior measures in a reproducible way. As an outlook aspects of the growing R&D topic "Research of protection against roadside bombing (IED)" are discussed.
*Keywords:   mine protection, blast loading, structural dynamics, occupant dynamics, dummy measurement.*

## 1   Introduction

In out-of-area and peacekeeping missions, mobility of armoured vehicles beside the cleared and swapped tracks is essential. In those situations the ground forces are endangered by different land mine types.

Since 1997, research institutes, universities and engineering companies in Germany have combined their expertise on protection of armoured vehicles in a working group. This group defined basic research programmes on blast and EFP protection and supports the MoD in transferring the resulting know-how in protection programmes of armoured land vehicles of the German Bundeswehr (fig. 1).

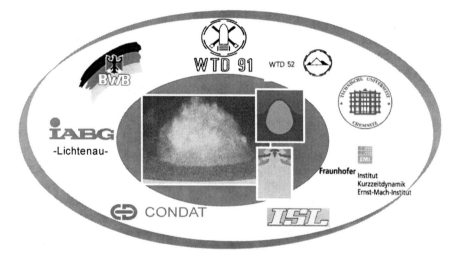

Figure 1:    Working group on landmine and IED protection.

The development processes of protection for armoured land vehicle systems started with reference data for mine effectiveness vs. mine protection effectiveness gained throughout basic plate trials. Real mines are classified by the amount of explosive contained in terms of equivalent TNT charge. For landmines filled with other explosives than TNT, the explosive charge is converted with available factors into TNT equivalent.

There are several parameters which affect the blast wave effects onto a vehicle structure:

The shape of the charge, the casing, the burial depth, the type of soil and in any case the standoff from the vehicle bottom (ground clearance) and the bottom plate mass.

An armoured vehicle basically is threatened by two different physical principles: by the explosive blast pressure and by kinetic energy projectiles (masses, fragments). Pressure mines are initiated under track and wheel, but there are also landmines with magnetic or tip wire detonators which initiate the charge below the belly. A mine protection solution has therefore not only to harden the attacked zone (bottom section) but also to design the complete compartment in a safe way. To handle the local mine effect the belly structure has to remain intact. This is a necessary first condition but local and global stresses migrate through the vehicle even if the blast can be kept outside.

High structural stresses will lead to local material collapse at fixation zones and interior devices and equipment can become a "secondary" projectile. Mine protection is a consideration of vehicle floor, crew cabin and occupant response.

Local deformations with velocities up to 200 m/s are a vital threat if they come in direct or indirect (for example via a seat structure) contact with the occupant. To meet these objectives the whole vehicle structure has to be analyzed to minimize local and global mine effects and therefore ensure the survivability of the vehicle crew.

An essential aspect in the development process of a mine protection solution is the numerical simulation.

The number of detonation trials in the development process is limited. Especially trials with dummy measurements are done only at milestone and qualification steps in the programme

A key aspect in the development of a mine protection system is therefore a combination of structural- and occupant simulation.

The calculation of the structural response under blast load starts with the pressure distribution under the bottom section in an Eulerian code. This blast model takes into account the type of explosive, the shape of the charge and soil conditions. It can be precalculated for different standoffs and then be mapped onto different structural FE-Models. The structural response at important zones of the vehicle model calculated via explicit codes (AUTODYN, LS-DYNA) can then be used as input for multibody occupant models (MADYMO).

Plate and segment tests which measure dynamic and plastic deformation and moreover accelerations at important structural zones have been used to calibrate material- and damage models.

A full spectrum of mine protection for a vehicle also takes overmatch situations into account. Numerical overmatch simulations show additional resources against higher threats than the system was designed for.

## 2　Aspects of structural dynamics

In mine protection the structural integrity of the bottom section respectively the crack resistance of the protection module is the minimum requirement. For an effective occupant protection in addition more boundary conditions have to be met. A direct contact of the occupant with highly stressed structural zones has to be avoided as those high shock accelerations cause severe traumatomechanic injuries.

All structural parts which are likely to come in contact to the occupants have to maintain within a tolerable acceleration level throughout the main stress phase.

A criterion of crucial importance is the dynamic bending directly above the detonation spot. If the local buckle with its high velocity which typically reaches up to 160 m/s within 1-2 ms comes in direct contact with interior structures the velocities will be transported to other components. This deformation is a parameter to be minimized during protection optimization. In case of a secondary contact the velocity of the dynamic bulge is a key parameter for the transport of structural stress transport. So if contact can't be avoided it has to be shifted to a "late" time point when the velocity already dropped into lower severity.

For the assessment of structural protection measures acceleration sensor data recordings are taken in each trial of a protection programme. They are measured with special high shock accelerometers. For an assessment of failure probabilities shock response spectra are plotted and frequency contents with high velocities and deformations are identified.

A measurement of structural accelerations directly above the detonation zone is hardly possible due to the damage high shocks above 100000 g cause even in damped sensors.

Instead of sensor data a high speed video of the dynamic bulge with a sufficient sampling rate (>10000 frames per second) allows the generation of dynamic motion data of marked tracers on even highest stressed zones in the video over a time duration of 1 to 20 ms. By means of pixel analysis of this high-speed video the dynamic bending of an arbitrary point visible during the deformation phase is calculated. A comparison of dynamic bending behaviour of different protection structures and an interpretation in categories of follow up damage potential is therefore possible. Figs. 2 and 3 show typical measurements during a mine detonation under a belly of a protected vehicle.

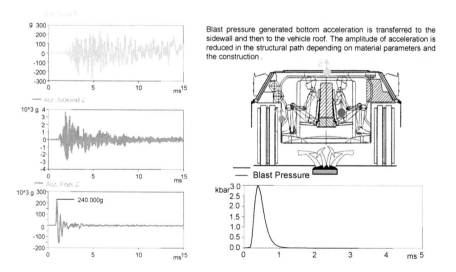

Figure 2:    Structural acceleration during a mine detonation under a vehicle belly.

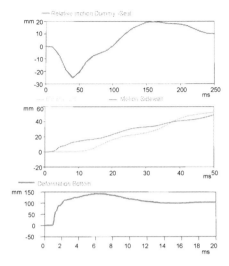

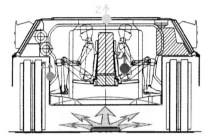

Due to the roof mounting of the seats the bottom deformation isn't directly transferred to the occupants. Relative to the seat the occupant is pushed into the cushion while the seat moves upward. This is followed by a rebound of the seat and the dummy is held back in the restraint system.

In general the roof deformation starts about 6-8 ms after onset of the sidewall deformation. Depending on the structure elastic roof movement can be greater than sidewall or floor movement.

Figure 3:    Structural deformation during a mine detonation under a vehicle belly.

Detonating the charge in a steel pit allows competitive testing without soil influence in a reproducible way, as described in STANAG 4569 (AEP-55 Volume 2). The configuration of steel-pit and charge is designed in such a way that the same local effects, i.e. bottom plate deformation and deformation velocity of a steel plate above the charge, were reached as for the situation of a reference mine surrogate buried into sand.

In early development phases plate trials on a test rig optimise the design of the protection module. Not only static values like plastic deformation but also the transient dynamics of a plate are taken into account.

## 3   Aspects of occupant dynamics

During a mine detonation underneath an armoured vehicle different effects lead to critical stresses onto the occupants.

Primary effects, like air pressure and splinters through bottom rupture, have to be prevented by the actual mine protection module. In addition the blast wave causes deformation in the bottom plate and shock migration through the connection elements and components which stand in direct contact to the occupant.

### 3.1 Dummy measurements

The development of protection measures is directed to reduce biomechanical loads during the detonation and rebound phase in a way that injury criteria limits are met. The measurement tool of choice is the Hybrid III 50% dummy-manikin. Although designed for frontal impact it's completeness of sensor locations at legs, pelvis, neck and head and its seating posture and biofidelity design made it fit for trials with different seat and restraint systems.

The dummy typically movement is filmed with 1000 to 2000 frames per second. The kinematic motion analysis allows in combination with the accelerometer values a balanced assessment of occurring stress levels. Remarkable events like collisions of femur or head with interior parts or even the roof construction can be identified.

In all German mine qualification programmes the HIII was the standard measurement tool for the acquisition of stress levels at certain locations combined with injury criteria. In improvised explosive device (IED) side facing impact situations the Euro SID dummies have proven to give good results since the starting point on injury criteria are based on ECE Norms and the Euro NCAP references. Experimental research and vehicular qualification trials with the HIII showed that certain body regions are most critical regarding acceleration based injuries:

In case of a mine detonation below the passenger compartment the crew is exposed to serious injury levels arising from:

- Feet/leg fractures/contusions by intruding bottom and pedal elements. (Feet and lower tibia which are most likely placed in endangered areas are loaded by high dynamic intrusions and bending.)
- Lumbar spine compression by high rate acceleration. (A transfer of accelerations into the seat system might lead to critical compressions in pelvis and lumbar spine.)
- Head and neck injuries by blunt impact onto rigid interior or roof parts: (an unrestrained occupant will hit interior parts or walls causing head accelerations and neck flexion and extensions above limit values).
- Injury caused by blunt and sharp impact of broken and loose equipment.

All channels of the dummy and the accelerometers on the structure usually are triggered by a CCD chip sensing the light flash of the ignition or a short circuit wire. The loading duration of a mine blast affected structure happens in a time duration shorter than 2-4 ms. This short impact causes the structure to respond according to its eigenfrequencies. The time duration of the maximal occupant loading usually takes from 15-40 ms. A measurement duration of 250 ms is in the majority of cases sufficient to make decisions on survivability of the occupants and equipment.

## 3.2 Injury criteria

In injury mechanics two important parameters determine the tolerance limit values of injury criteria [1, 2]:

- injury probability and
- injury severity.

In STANAG 4569 mine protection testing guidelines most nations agreed to a tolerable severity level of AIS 2+ connected with a risk probability of 10%. This means in common sense that occupants are capable of leaving the vehicle on their own after a mine detonation.

In NATO/RTO group HFM-090/TG25 the injury criteria were discussed and harmonized for STANAG 4569, AEP 55 Volume 2.

## 4 Basic research

### 4.1 Test Rig for Occupant Safety Systems TROSS®

Main focus is the response of structures under blast loading conditions and how to get information about local accelerations acting on the occupant. Interior systems like footrests, pedals and seats transport the acceleration with different transfer properties. Detonation tests within the TROSS experiments analyse stress reduction potential by decoupled designs and optimised restraint systems.
As a worst case benchmark rigid steel seats and rigidly fixed footrests are evaluated e.g. against up to date air cushioned truck seats or new designed crashworthy connection elements (fig. 4).

Steelchair Trial:                              Steelchair MADYMO Simulation:

 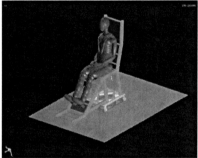

Figure 4:     Steel chair dummy test and occupant simulation.

### 4.2 TROSS I® test campaign example "interior collision"

Due to the limited space in military vehicles the problem of head contact with roof structures and femur steering wheel contact was investigated with the TROSS I test rig. In the scaled charge trials typical mine detonation footrest and seat input velocities were reached by scaling down the charge and reducing the standoff. Bottom Plate deformations of up to 100 mm brought the HIII head in hard contact to a rigid roof plate (fig. 5).

Maximum acceleration in case of head contact could clearly be reduced by a helmet but on the other hand neck loading duration was extended (fig. 6). Wearing a helmet in case of no head contact had hardly any influence on neck forces and moments at all.

Figure 5:    Test configuration for steering wheel/femur and head/roof contact.

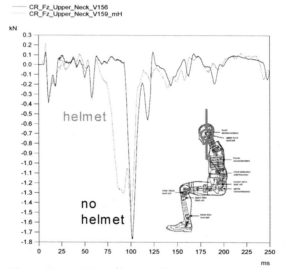

Figure 6:    Exemplary neck compression force.

## 4.3 TROSS II® test campaign example "sand/steel pit comparisons"

The scaled tests in the TROSS I test rig remained in the elastic plate bending area. To simulate high structural shocks not only in the bottom plate but also in the whole compartment enhanced test rig trials were done. A free moving cabin in the weight class of light armoured vehicles allows testing of all kinds of mission important systems with special focus on occupant stresses transferred by seating and footrest parts (fig. 7). The lower limit for the dummy movement within the cabin showed to be the global movement although local velocities are above this level.

Structural velocities of sidewall or roof reach more than 10 m/s within few milliseconds whereas the dummy velocity and the global velocity only reach 6 m/s and 4 m/s respectively (figs. 8 and 9).

Figure 7:     TROSS II tests in sand.

**time table of blastmine detonation including global movement**

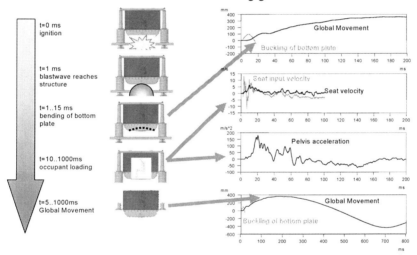

Figure 8:     Measurements at TROSS II test rig blown up by a 1:1 charge.

## 5   Conclusions

Contact trials showed helmet effects to be ambivalent. Kevlar helmets reduce head space and increase neck force duration. Only in case of sharp edge contact positive acceleration reduction effects outbalanced the negative force increasing influences.

Test campaigns with charges out of sand and out of a steel pit showed equal results in short time structural dynamics and occupant loading. Differences in impulse transfer were shown to be not relevant for dynamic bottom bending and velocities. Compared to local maximum cabin deformations the global movement in general was shown to be of minor importance for injury criteria values.

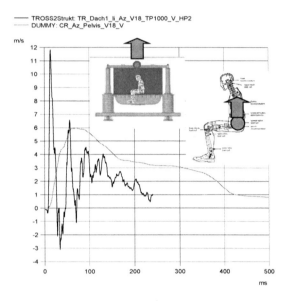

Figure 9:     Structural velocity and dummy pelvis velocity.

# 6  Outlook

The test rig approach which has been proven successfully in the field of mine protection is now adopted for the terrorist IED threats (fig. 10).

Forming a testing matrix with relevant charges and standoffs and material testing is started by means of dynamic plate and segment tests.

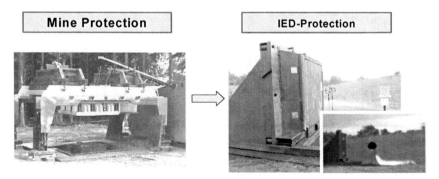

Figure 10:     Test Rigs for Mine- and IED protection.

Based on the integrated test rig TRAILER®, which allows to study the dynamics and forces in highly stressed structures directly above a detonating mine, IABG has developed a new test environment called IED-TRAILER® for systematic investigations on the effects of large blast IEDs.

In a first experimental series, the dynamic response of different steel plates exposed to high blast load has been examined. The HE charges varied from 2 kg to 300 kg TNT at detonation distances between 2 m and 15 m. Pressure, momentum and time resolved bulging of the plates have been measured. The results provide a basis for the characterization of the severity of different blasting conditions.

To facilitate test series with large charges, IABG has worked out a method by which the same response of a structure as achieved by a large IED can be generated with a drastically reduced amount of explosive. The method has been validated and its range of application determined. Actually activities are running to extend the IED testing on vehicle segments and complete vehicles.

## Acknowledgement

The results presented in the paper are an IABG Lichtenau part of the R&D Programme of the German working group on mine protection of armoured vehicles coordinated and funded by WTD 91 Dept. 410 Meppen proving ground.

## References

[1]     Mertz, H., Anthropometric test devices. *Accidental Injury Biomechanics and Prevention*, eds. A.M. Nahum & J.W. Melvin, Springer-Verlag: New York, pp. 89-102, 2002.

[2]     King, A.I., Fundamentals of impact biomechanics: part 2—Biomechanics of the abdomen, pelvis, and lower extremities. *Annual Review of Biomedical Engineering*, **3**, pp. 27-55, 2001.

# Blast testing of CFRP and SRP strengthened RC columns

J. O. Berger, P. J. Heffernan & R. G. Wight
*Department of Civil Engineering, Royal Military College of Canada, Canada*

## Abstract

Blast testing on scaled reinforced concrete columns was conducted to study the behaviour of Steel Reinforced Polymer (SRP) wrapped columns in both flexure and shear. Two vertical testing frames were constructed to support each two columns per blast in a fixed-fixed configuration while applying a static axial load of 300 kN. The specimens were exposed to blast waves at a variety of incident pressures which resulted in damage from minor to severe. A reflected impulse which resulted in moderate damage was then selected and used to study the effects of varying the density of SRP wraps. CFRP strengthening was also used in order to compare the effects of the two strengthening materials. An SRP/CFRP hybrid combination using SRP for longitudinal or flexural strengthening and CFRP sheets for transverse or shear strengthening was tested. It was observed that the SRP strengthened columns were quite similar to those strengthened with CFRP. The experimental results were compared to both analytical SDOF models as well as numerical models created using advanced explicit analysis software. SRP appeared to be a very effective external strengthening material for increasing the resistance of concrete components, providing similar performance to other FRP (CFRP) wraps at potentially lower cost. The SRP materials proved to be quite resilient even when exposed to a close-proximity explosion where spalling of the RC columns was significantly reduced.

*Keywords: blast, reinforced concrete columns, SRP, CFRP, strengthening.*

## 1 Background

During the past decade, significant research has been carried out on the strengthening of reinforced concrete (RC) slabs, beams and columns using externally bonded carbon fibre reinforced polymer (CFRP) sheets. CFRP sheets bonded to the surface of concrete members have been found to be an efficient and effective strengthening method that requires little heavy equipment or manpower to install but can significantly enhance their strength and serviceability. CFRP sheets have been proposed as a method of strengthening concrete and masonry components against the effects of blast. Initial tests show that they may be very appropriate for some blast strengthening situations [1]. Steel reinforced polymer (SRP) sheets have recently been proposed as an alternative to CFRP to strengthen reinforced concrete beams. Initial tests with the material applied to concrete beams indicates that SRP appears to have a much higher lateral shear strength and is much tougher than CFRP sheets [2]. Because of this property, SRP sheets may be more suitable than CFRP sheets for strengthening concrete components subject to blast. Early testing using SRP as transverse or shear strengthening for scaled columns [3] indicated both good structural performance of the wraps as well as high resilience when exposed to a near field explosion.

## 2  Experimental study

### 2.1  Concrete specimens

To determine the suitability of SRP sheets for strengthening RC members against the effects of blast loading, a series of destructive tests were conducted using explosives and scaled RC columns strengthened with various configurations of SRP and CFRP external strengthening. This program builds on the lessons learned from previous blast testing of SRP strengthened RC members [4]. Transverse and longitudinal cross-section profiles of these columns are shown in figs. 1 and 2, respectively. The concrete specimens were 150 mm by 150 mm in cross-section. The total length of the specimens was 2100 mm and the unsupported length during testing was 1500 mm. The section was reinforced with four 10 mm diameter reinforcing bars and 6 mm ties spaced at 100 mm. The 28-day compressive strength of the concrete was 44 MPa, the yield strength of the internal reinforcing steel was 450 MPa and the ultimate strength was 630 MPa.

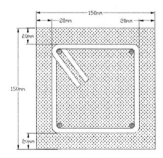

Figure 1:     Transverse cross-section of RC specimen. Carrière (2004).

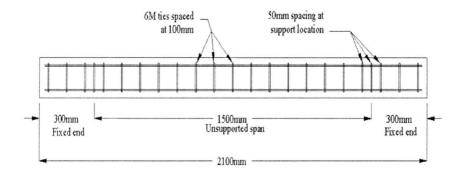

Figure 2:     Longitudinal cross-section of RC specimen.

## 2.2  Specimen strengthening with CFRP and SRP

Specimens were strengthened with either MBrace® CF130 CFRP sheets or Hardwire® 3x2 SRP sheets in either the transverse and/or longitudinal direction. 300 mm at each end of each column specimen was strengthened using SRP sheets, where the columns were fixed into the testing frame, regardless of configuration. Specimen strengthening configurations were selected to facilitate several comparisons. These comparisons include varying the blast intensity for SRP strengthened (longitudinal and transverse) columns, varying the quantity of longitudinal SRP strengthening at a constant blast load, comparing the effect of longitudinal versus transverse strengthening, and finally comparing the effectiveness of CFRP v. SRP materials for strengthening in either the longitudinal or transverse direction. Numerous control columns were also tested. Specimen configurations are given in table 1. Longitudinal strengthening, when applied, was only bonded to the side of the column opposite the explosive charge.

For this project, a high density 300 mm wide SRP tape was selected, with a 3x2 chord type and a protective brass coating. The epoxy used to adhere the SRP to the RC specimens was SikaDur® 330, a two-component, moisture-tolerant, high-strength, and high-modulus epoxy resin. Ancillary testing on SRP specimens [2] indicated that the manufacturer's specified modulus and ultimate strength of 77.9 GPa and 1170 MPa respectively were representative of the material. These values were used for this project. CFRP strengthening materials used were MBrace® CF130 CFRP sheets. Ancillary testing on CFRP specimens carried out in prior projects indicated that the CFRP had a fibre modulus and rupture strain of 228 GPa and 1.67% respectively.

The initial step in fabricating the specimens was to prepare the concrete surfaces. Initially, the longitudinal sheet of 1500 mm was applied to the specimens. The width of the SRP sheet was 120 mm while the width of the CFRP sheet was 40.1 mm in order to achieve similar stiffness as the SRP. For the specimens wrapped with SRP, the SRP was first bent into square-shaped strips using a sheet metal bender. The length of each side of the square strip was

2 mm greater than the 150 mm width of the RC members. Since the strips were cut to 700 mm lengths prior to bending, this resulted in a 92 mm overlap of the material. A 1 to 2 mm layer of epoxy was spread on the surface of the concrete member before the SRP strips were applied. Several wraps of copper wire were used to hold each SRP strip in place during curing. A final layer of epoxy was applied using the copper wire as a thickness gauge for the final layer. The RC member was clamped with strips of plywood that were wrapped in sheets of plastic and applied to all four sides. After curing for twenty four hours, the clamps and plywood were removed. All wrapped specimen dried for at least a week before being tested.

The methods used to strengthen the columns with CFRP sheets were similar to those used for SRP. For the columns with CFRP transverse reinforcement it was necessary to round the corners of the RC Columns to a radius of .02 m to avoid a premature failure of the CFRP.

Table 1: Specimen configuration and blast parameters.

| | Specimen | Strengthening | | Charge Weight (kg C4) | Range (m) |
| | | Longitudinal (% SRP K) | Transverse ? | | |
|---|---|---|---|---|---|
| Control | A1 | nil | nil | 65 | 4 |
| | B1 | nil | nil | 65 | 5 |
| | C1 | nil | nil | 100 | 4 |
| | D1 | nil | nil | 100 | 5.25 |
| | I1 | nil | nil | 49* | 4 |
| SRP | A2 | SRP | SRP | 65 | 4 |
| | B2 | SRP | SRP | 65 | 5 |
| | C2 | SRP | SRP | 100 | 4 |
| | D2 | SRP | SRP | 100 | 5.25 |
| | I2 | SRP | SRP | 49* | 4 |
| % SRP | F2 | SRP | SRP | 90 | 4.5 |
| | E1 | SRP (75%) | SRP | 90 | 4.5 |
| | E2 | SRP (50%) | SRP | 90 | 4.5 |
| L v. T | G2 | nil | SRP | 100 | 5.25 |
| | G1 | SRP | nil | 100 | 5.25 |
| CFRP | F1 | SRP | CFRP | 90 | 4.5 |
| | H1 | CFRP | CFRP | 100 | 4 |
| | H2 | CFRP (75%) | CFRP | 100 | 4 |

Notes: K is equivalent stiffness of SRP application. No number implies 100%.
    *Hemispherical charge.

## 2.3 Testing facility

Testing was conducted on a Canadian military establishment. Labour support for the testing program was provided by the Canadian Forces' 2 Combat

Engineer Regiment. To hold the RC specimens in place during the explosive testing, two steel vertical test frames were designed and constructed. Each frame was anchored and counter-balanced to resist the blast loading. The testing frame restrained a 300 mm length at each end of the specimen, resulting in a fixed-ended boundary condition.

To simulate column behaviour, it was necessary to apply an axial load to the specimens during blast loading. This was applied as a static load prior to each test. Hydraulic jacks were used to apply a 300 kN per column vertical compressive load to the extensible frame that constrained the columns. Threaded rods running through the frame were tightened into position and sustained the load at the start of the test. The jacks were removed prior to testing.

To facilitate the application of the blast loading, the explosive charge was placed on a platform at mid-column height a specified distance from the face of the column. Distances to the charge were measured to the centre of gravity of the charge. With two supporting frames, a maximum of 4 columns at 2 different distances were tested per explosive charge.

## 2.4 Testing program

The intent of this testing program was to examine the performance of various strengthening schemes, as noted in Table 1. Blast loads were applied that would cause moderate damage in the strengthened specimens, but were likely to heavily damage an unstrengthened column. As an ideal, it was preferred to increase both distance and charge size for equivalent blasts so as to approach planar blast waves and minimize fireball effects. Maximum charge size limitations and resource allocations also affected the blast load selections. The program planned with spherical charges but due to the softness of the ground, hemispherical charges had to be used.

## 2.5 Instrumentation

Instrumentation was mounted on the specimens and in the free-field in the vicinity of the explosions. Instrumentation included strain gauges, accelerometers and pressure transducers. High speed video also captured the explosion. The blast was initiated from, and all instrumentation and monitoring personnel were located in, an Armoured Data Acquisition Vehicle (ADAV), located approximately 100 m from the blast location. Four strain gauges were mounted on the reinforcement inside the concrete members to monitor any deformation inside the RC members during the residual strength testing. Two gauges were located at mid height of the column while two others were situated near the anticipated plastic hinge location at the support. Accelerometers were monitored with a WaveBook/516E™ 1 MHz Ethernet based portable high-speed waveform data acquisition system. Free field blast pressures were collected using five 34 MPa (5000 psi) PCB™ Free Field Blast Pressure Probes. These probes were used for each test and positioned at various distances and at right angles from the blast within the free field. Two probes were relocated for each test to be at the same distance from the charge as the column specimens in the two frames. The probes were positioned at a height of 1.5 m off the ground. This height was selected to avoid disturbances to the blast wave resulting from surface effects. Readings from the probes were made using an MREL® DataTrapII high speed data recorder that recorded channels at 5 MHz per channel. Some columns were affixed with a Model 52 accelerometer by Measurement Specialties Inc. capable of measuring 2000 g at 7 kHz. The accelerometers were affixed to the rear face at mid-height of the columns. High speed video of the explosions was recorded using an Olympus i-Speed video camera recording at 1000 frames per second (fps). All data acquisition units and the camera were initiated by trigger wired mounted around the explosive. C-4 explosive was arranged into either a square or hemispherical bundle for each test and was electronically detonated with the initiation originating at the centre of the bundle. The charge was placed on a wooden platform at the mid-height of the columns as shown in fig. 3.

Figure 3:     Experimental setup in the field.

# 3   Experimental results

Summary results of each specimen tested are given in table 2. Rotations are given at mid-height of the column based on the plastic deformation and mid-height displacements that were observed. Rotations are for the final static deformed shape of the column. Instrumentation difficulties with some of the accelerometers precluded reasonable calculation of the maximum dynamic deflections.

### 3.1 SRP and CFRP subjected to blast

For all tests the specimens were located between 4 and 5.25 m from the COG of the explosive. This caused each specimen to be exposed, in succession, to the high-pressure blast wave followed by the heat of the afterburn. Further, on inspection of the specimens, it was apparent that small debris objects acted as secondary projectiles and struck the specimens in numerous spots. Fig. 4 shows some of the damage experienced by the SRP wrapped specimens. Areas where the epoxy resin has superficially delaminated are apparent as are small isolated damage areas due to the impact of debris projected during the blast. These damage areas tend to be localized and of a superficial nature.

Figure 4:     Blast Damage to SRP resulting from 49 kg blast. Some impact damage from debris was evident.

CFRP specimens also suffered damage to the laminate dependent on the magnitude of the blast experienced. Pieces of CFRP sheets were torn off of the columns in several places ranging from small to modest sized (150 mm). At equivalent distance and charge size, the SRP appeared to suffer less surface damage than the CFRP.

### 3.2 Reinforced concrete columns subjected to blast

The results of the RC columns subjected to blast are given in table 2. Charge sizes were slowly increased during the early tests until a complete failure of the unstrengthened RC column was completely destroyed. Total destruction of the unstrengthend RC column occurred at a blast of 100 kg (C4) at a range of 4 m with significant damage and end rotation apparent in the specimen tested at 5.25 m at a charge of 100 kg. These related to peak overpressures of 1930 kPa and 1327 kPa for the partially damaged and fully damaged columns respectively. The strengthened columns were tested in this range in order to denote improvements relative to the destroyed control columns. (For all tests, the frame was not fixed to the ground so the columns were not exposed to the full effect of the pressure nor the impulse).

### 3.3 SRP Strengthened reinforced concrete columns subjected to blast

Columns fully strengthened with SRP in both the longitudinal and transverse directions were paired with the various control beams previously described. As would be expected, where insignificant damage to the unstrengthened columns was noted, the SRP strengthened columns were also undamaged. In the instance of the severely damaged control beam (4.6° mid-height rotation) its paired strengthened beam was only damaged slightly (1.2° mid-height rotation). In the case of the totally destroyed unstrengthened column, its matched SRP column suffered major deformations (6.3° mid-height rotation), including rupture of the SRP longitudinal strengthening but otherwise remained intact and able to carry significant residual load. Rupture of the SRP strengthening of Column C2 post residual axial testing is shown in fig. 5.

For the series of columns with varying amount of longitudinal SRP, only slight damage was noted with little permanent deformation in the columns.

For the two columns which were strengthened with either longitudinal or transverse SRP, both exhibited minor permanent deflections (0.4° and 0.3°, respectively) of roughly the same magnitudes. Permanent cracks, delamination and spalling were evident in the specimen strengthened with longitudinal SRP only. The control beam tested at the same reflected impulse suffered major damage (4.6°) though it survived the blast. Notably, the column strengthened with SRP in both the longitudinal and transverse directions and loaded at the same reflected impulse had slightly higher permanent deflections (1.2°) than the columns strengthened with one or the other type of strengthening.

Table 2:     Specimen results following blast.

| | Specimen | Scaled distance Z | Rotation at mid height (degree) | Remarks |
|---|---|---|---|---|
| Control | A1 | 0.94 | 0 | Intact |
| | B1 | 1.18 | 0 | Intact |
| | C1 | 0.82 | N/A | Total destruction |
| | D1 | 1.07 | 4.6 | Cracks (up to 3mm), spalling |
| | I1 | 1.03 | 0 | Intact |
| SRP | A2 | 0.94 | 0 | Intact |
| | B2 | 1.18 | 0 | Intact |
| | C2 | 0.82 | 6.3 | Compression bulge, longitudinal SRP rupture |
| | D2 | 1.07 | 1.2 | Cracks (up to 0.5mm) |
| | I2 | 1.03 | 0 | Intact |
| % SRP | F2 | 0.95 | 0 | Intact |
| | E1 | 0.95 | 0.9 | Crack (1mm) |
| | E2 | 0.95 | 0 | Intact |
| L v. T | G2 | 1.07 | 0.3 | Cracks (up to 0.35mm) |
| | G1 | 1.07 | 0.4 | Cracks, de-lamination, spalling |
| CFRP | F1 | 0.95 | 0 | Small pieces of CFRP torn off (up to 3 cm long) |
| | H1 | 0.82 | 1.5 | Cracks, small pieces of CFRP torn off, longitudinal CFRP rupture |
| | H2 | 0.82 | 1.3 | Pieces of CFRP torn off (up to 15 cm long), longitudinal CFRP rupture |

Applying SRP on RC columns required much less effort than the application of CFRP if the recommended surface preparation and rounding off of the corner for the CFRP application is considered.

### 3.4 CFRP strengthened reinforced concrete columns subjected to blast

The CFRP strengthened specimens (H1 and H2) tested at the equivalent charge/distance as an SRP strengthened column (C2) had less plastic deformation. The transverse reinforcement of the CFRP columns was 3 times the stiffness of the equivalent SRP strengthened column. This resulted from the desire not to leave gaps in the transverse strengthening. It was accepted that the transverse CFRP strengthened columns would be stiffer. The CFRP strengthened columns were bent locally at the plastic hinges but remained largely straight between the plastic hinges. Conversely, the SRP strengthened column had a distributed curvature along its full length.

Figure 5:     Column C2 post blast and post residual axial testing. Some transverse strengthening removed.

## 4   Conclusions

The following conclusions are deduced from the experimental results:
- Significantly less damage was observed in columns strengthened with longitudinal and transverse SRP then in un-strengthened specimens
- Columns strengthened with SRP appeared to be more ductile the those reinforced with CFRP
- SRP appeared to better resist small projectile impact then CFRP

## Acknowledgements

The authors would like to acknowledge the financial and practical support of the Department of National Defence through the Military Engineering Research Group and the support of the Canadian Forces' 2 Combat Engineer Regiment. Further financial support in this area by Intelligent Sensing for Innovative Structures (ISIS) Canada Research Network is appreciated.

# References

[1] Crawford, J., Malvar, J., Morrill, K. & Ferrito, J., Composite retrofits to increase the blast resistance of reinforced concrete buildings. *Symposium on Interaction of the Effects of Munitions with Structures*, 2001.

[2] Prentice, D.B. & Wight, R.G., Prestressed SRP and CFRP sheets for strengthening concrete beams. *Third International Conference on FRP Composites in Civil Engineering (CICE 2006)*, Miami, Florida, USA, 2006.

[3] Carriere, M., Wight, R.G. & Heffernan, P.J., Effect of blast loads on reinforced concrete strengthened with steel reinforced polymer sheets. *CD-ROM Proceedings of the 8th International Symposium on Fiber-Reinforced Polymer Reinforcement for Concrete Structures (FRPRCS-8)*, Patras, Greece, 2007.

[4] Carriere, M.D., *Steel Reinforced Polymer Strengthening of Reinforced Concrete to resist Blast Loads*, Royal Military College of Canada, 2006.

# Analysis of the explosive loading of open-ended steel pipes

N. Rushton[1], G. Schleyer[1] & R. Cheesman[2]
[1]*Department of Engineering, University of Liverpool, UK*
[2]*Atomic Weapons Establishment (AWE), UK*

## Abstract

A programme of numerical, theoretical and experimental studies has been carried out at the University of Liverpool on the explosive loading of seamless steel pipes. Ten pipes of dimensions 800 mm long, 324 mm outside diameter and 9.5 mm thick were subjected to explosive loading along their mid-length using cylindrical PE4 charges detonated simultaneously from both circular faces. A range of charge sizes were used to determine the effect of the blast waves on the maximum plastic hoop strain in the pipe walls with the aim of determining the magnitude of the impulse required to initiate failure in the pipe wall. Theoretical analyses of the deformation process were investigated using equations available in the past literature as well as an attempted derivation of equations of motion used to analyse the transient deformation of the pipe where wall thinning is considered based on a constant volume assumption. A comparison of these analyses is made in this paper together with a numerical analysis of the problem using AUTODYN 2D where a von Mises material model was used to simulate the assumption of a perfectly plastic material. Further finite element simulations were conducted to model the blast process and structural response of the experimental pipes using cylindrical charges. It was found that a Johnson-Cook strength model for the simulated pipe material in the numerical solution gives good agreement with the test data as this model attempts to account for the strain rate and strain hardening effects of the steel. The objective of the study, which is sponsored by AWE plc, Aldermaston, is to ultimately determine the failure mechanism of such a pipe under very high rates of loading.

*Keywords: blast loading, seamless steel pipes, numerical simulations.*

# 1 Introduction

The detonation of an explosive releases a sudden burst of energy, which results in the propagation of a shock wave through the explosive material. The duration of these shock waves are of the order of microseconds and are characterised by a peak overpressure, $P_0$, followed by an almost exponential decay in pressure as the shock front passes. If the detonation occurs in a confined space the typical pressure-time history for a high explosive, such as PE4, is illustrated in fig. 1 where the pressure wave is reflected off the confining structure to produce successive peaks of decreasing magnitude.

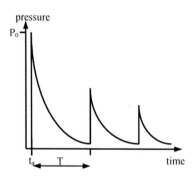

Figure 1:    A typical high explosive pressure-time history for a confined detonation.

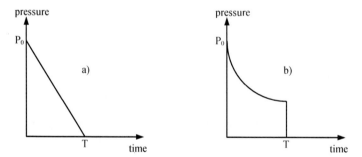

Figure 2:    Pressure-time history idealisations: a) triangular decay; b) exponential decay.

Analytical methods for determining the response of structures to blast loading are usually concerned with the initial pulse duration, T, as this is of a far higher impulse magnitude than the successive pulses. Typical idealisations of this pulse are the triangular decay, exponential decay shown in fig. 2.

When a containing structure, such as a cylinder, is subjected to the explosive loading by a detonated charge it can undergo substantial plastic deformation up to a maximum strain upon which subsequent motion is elastic involving both radial oscillations and other vibrations. Whereas the loading impulse duration lasts for microseconds, the wall response takes milliseconds to occur. This large difference between explosion loading time and vessel response time defines the loading imparted on the vessel wall as being impulsive.

## 2  Theoretical analyses

The response of cylinders to blast loading has been studied by a few authors and equations have been produced to predict the final hoop strains or radial displacements of the vessel walls. Duffey and Mitchell [1] based their analysis on a rigid-perfectly plastic material loaded by a centrally placed spherical charge. Assuming the material is not sensitive to strain hardening or rate effects and the pressure pulse is idealised as triangular, the equation predicting the final hoop strain ($\varepsilon_{\theta(max)}$) was found to be

$$\varepsilon_{\theta(max)} = \frac{I^2}{2\rho\sigma_y h^2},$$  (1)

where I is the impulse, $\rho$ is the steel density, $\sigma_y$ is the yield stress and h is the wall thickness.

A simplified analysis for an impulsively loaded cylindrical shell was produced by Fanous and Lowell [2] in which a single-degree-of-freedom (SDOF) analysis was combined with energy methods to predict the final wall displacements. The material was assumed to be ductile with an elastic-perfectly plastic material model. Other assumptions included a localised impulsive loading and assumed displacement profile to produce the following expression for the final maximum hoop strain

$$\varepsilon_{\theta(max)} = \frac{3i^2}{8M h\sigma_y} + \frac{\varepsilon_y}{2},$$  (2)

where i is the specific impulse, $\varepsilon_y$ is the yield strain and M is the mass per unit area of the pipe.

A more recent paper was produced by Clayton [3] on the design of vessels for explosion containment. A SDOF elastic analysis was used to determine dynamic magnification factors for the response of cylindrical or spherical vessels depending on the natural frequency of the structure. Although this paper is focussed on the design of structures to resist explosions without yielding it has been shown that rule-of-thumb analyses of the stresses can be used to determine the maximum plastic deformations for impulsively loaded cylinders using the equation

$$\varepsilon_p = \frac{\sigma_e \varepsilon_e}{\sigma_p}.$$  (3)

where $\varepsilon_p$ is the plastic strain, $\sigma_e$ is the elastic stress, $\varepsilon_e$ is the elastic strain and $\sigma_p$ is the plastic stress.

Since none of the equations mentioned in this section consider the effects of wall thinning on the transient deformation of the shell, equations of motion have been derived based on a similar analysis done by Baker [4] for spherical shells. This analysis uses an incremental approach to the SDOF problem by analysing the wall displacements at different times during the deformation process. Starting from first principles, an element of the cylinder wall is analysed to determine the equations of motion as seen in fig. 3, where $\sigma_\theta$ is the hoop stress, $\sigma_L$ is the longitudinal stress, $R_i$ is the inner radius of the shell and $u_r$ is the radial displacement at any time, t.

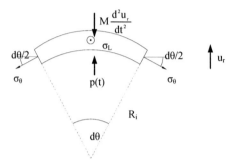

Figure 3:    Element of a cylindrical shell under transient loading.

During the initial phase of motion, where the wall displacements are elastic and a triangular pressure pulse is assumed, the equation of motion for the cylinder wall is

$$\frac{d^2 u_r}{dt^2} + \omega^2 u_r = \frac{p(t)}{\rho h}, \tag{4}$$

where the circular natural frequency, $\omega$, of the cylinder is given as

$$\omega = \sqrt{\frac{E}{\rho\left(1 - v^2\right)R_i^2}}, \tag{5}$$

where E is Young's modulus and $\upsilon$ is Poisson's ratio.  Solving eqn (5) and applying suitable boundary conditions gives the transient radial displacement as

$$u_r = \frac{P_0}{\omega^2 \rho h}\left(\frac{\sin \omega t}{\omega T} - \cos \omega t + 1 - \frac{t}{T}\right). \tag{6}$$

If it is assumed that the material behaves with an elastic-perfect plastic model, after the material has yielded the equation of motion becomes

$$\frac{d^2 u_r}{dt^2} + \frac{\sigma_y}{\rho R_i} = \frac{p(t)}{\rho h}. \tag{7}$$

This equation is solved to give the following expression for the plastic pipe displacement

$$u_r = u_{ry} + \dot{u}_{ry}(t-t_y) + \frac{P_0}{\rho h}\left(\frac{t^2}{2} + \frac{t_y^2}{2} - t_y t\right) + \frac{P_0}{\rho hT}\left(\frac{t_y^2 t}{2} + \frac{t^3}{6} - \frac{t_y^3}{3}\right), \qquad (8)$$

$$+ \frac{\sigma_y}{\rho R_i}\left(t_y t - \frac{t_y^2}{2} - \frac{t^2}{2}\right) \qquad\qquad 0 \le t \le t_y$$

Eqn (8) applies until the pulse duration has ended after which the equation of motion becomes

$$u_r = u_{rT} + \dot{u}_{rT}(t-T) + \frac{\sigma_y}{\rho R_i}\left(Tt - \frac{t^2}{2} - \frac{T^2}{2}\right) \qquad t \ge T. \qquad (9)$$

The final wall displacement is found when the shell velocity reduces to zero. Wall thinning is accounted for at each time interval using a constant volume assumption where the incremental thickness, $h_2$, and radius, $R_{i2}$, are related to the initial values by the following quadratic

$$h_2 = \frac{-2R_{i2} + \sqrt{4R_{i2}^2 + 8R_i h + 4h^2}}{2}. \qquad (10)$$

## 3  Experimental investigation

### 3.1  Material characterisation

Two 6 m lengths of API 5LX-42 seamless mild steel pipes were sectioned to form cylinders of lengths 0.2 m and 0.8 m for material characterisation tests and blast tests respectively. To date only static tension tests have been performed on the pipe material. Due to the lack of sufficient material through the pipe wall thickness, specimens could not be manufactured to measure stresses and strains in the hoop direction so only longitudinal values could were obtained. The test results for the material in tension are presented in Table 1 as a range of values.

Table 1:  Static tensile material properties for the API 5LX-42 pipe material.

| | | |
|---|---|---|
| Yield Stress | 289.4 – 344.2 | MPa |
| UTS | 518.7 – 511.7 | MPa |
| Young's Modulus | 192.2 – 203.1 | GPa |
| Poisson's Ratio | 0.28 – 0.29 | |

### 3.2  Blast tests

A series of tests were conducted on the 0.8 m long, 324 mm diameter, 9.5 mm thick open-ended steel pipes using centrally loaded, cylindrical PE4 charges of different masses as shown in fig. 4. The charges were detonated simultaneously from both circular faces causing the propagation and collision of the shock waves at the centre of the explosive to impart a radial loading on the pipe mid-length. In order to prevent any constraint to the deformation of the wall, the pipes were supported horizontally using a combination of a wooden trestle and slings. Aligning the pipe to the horizontal served to minimise the effects of reflected pressure waves from nearby structures.

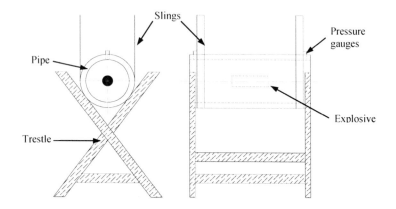

Figure 4:     Experimental setup for the blast loading of steel pipes.

Tests started with 0.6 kg of PE4 with the mass increasing for different pipes in order to try and attain the mass required to initiate pipe wall failure. During each test the transient pressure was recorded using a pair of Kulite pressure gauges placed 100 mm from the open ends of the pipes where they would not be able to influence the deformation or failure process. The maximum hoop strain was measured at the end of each test, transient strains not being able to be recorded due to the high wall accelerations causing any attached strain gauges to debond.

## 4    Numerical analysis

The blast loading of the pipes was modelled using AUTODYN 2D [5]. Due to the axially symmetric nature of the problem the 2D tool was used to reduce computation time for each analysis. Two strength models were used to characterise the pipe material, namely von Mises and Johnson-Cook [6] for the comparison of the numerical results with the analytical and experimental ones respectively. It has been shown that the Johnson-Cook model is well suited for taking into account the combination of thermal softening due to high strain rate loading and strain hardening in the material, the form of the equation being given as

$$\sigma = \left[ A + B\varepsilon^n \right] \left[ 1 + C \ln \dot{\varepsilon}^* \right] \left[ 1 - T^{*m} \right], \tag{11}$$

where A is the yield stress, B is the strain hardening constant, n is the strain hardening exponent at a strain rate of 1 s$^{-1}$, C is the strain rate constant and the final bracketed term gives an expression to account for thermal softening of the material. All constants are determined experimentally from tensile test data on the pipe material. However, as only static tests have been conducted to date and there is no data available for class API 5LX-42 steel, values for the constants were taken from those of 4340 steel as obtained by Johnson and Cook [6].

The pipe material was modelled as a Lagrange mesh within the Euler mesh used to represent the surrounding air and PE4 explosive. The PE4 explosive was modelled using the JWL equation of state. The setup for the simulation can be seen in fig. 5. This setup uses a spherical PE4 charge detonated from the centre for the purpose of comparing numerical and theoretical results. The PE4 charge is modelled as a cylindrical structure for modelling the experimental setup.

Figure 5:    AUTODYN 2D setup of an axi-symmetric cylindrical shell loaded with a spherical PE4 charge.

## 5   Results

### 5.1 Theoretical and numerical comparisons

Approximate analyses on the deformation of open-ended steel pipes under impulsive loading were performed using the equations discussed in Section 2. In each of these analyses the explosive charge was assumed to be spherical and centrally detonated with values of the peak pressures and impulses for each charge obtained from report TM5-1300 [7] on TNT free air blast data with the relative effectiveness factor of 1.33 applied to model PE4. The other parameters used in the equations were determined from nominal pipe material properties and dimensions. The results of these theoretical analyses together with the results of numerical simulations of the problem can be seen in fig. 6.

It can be seen in fig. 6 that there are close approximations in hoop strains between all the analyses with the exception of that proposed by Fanous and Lowell [2] for explosive masses up to 0.6 kg. For charge weights in excess of this value, the proposed SDOF model shows that the numerical results lie between the strains determined when wall thinning in considered and omitted, the differences between the values increasing as the mass of PE4 increases.

Out of the three theoretical analyses found in the literature, the equation proposed by Duffey and Mitchell [1] is the best approximation of the numerical solution although the difference in strain values increases with increasing charge weight. The equation proposed by Fanous and Lowell [2] consistently predicts lower strain values than the other analyses up to a charge weight approximately 1 kg where the Clayton [3] analysis predicts the lowest strains per mass of PE4.

For spherical charge sizes less than 0.6 kg similar predictions for the final plastic hoop strain can be found using any of the theoretical analyses mentioned in Section 2 with the exception of the Fanous and Lowell [2] analysis. This analysis was not entirely suited to the selected problem as it is based on the assumption that the explosive is localised to a particular section of the cylinder wall where an assumed deformation profile is used.

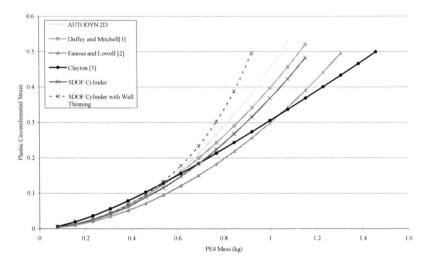

Figure 6:    Comparisons between the theoretical and numerical analyses on the maximum hoop strain in a blast loaded cylindrical shell.

## 5.2 Experimental and numerical comparisons

Out of the nine blast tests conducted to date, six of the pipes underwent substantial plastic deformation and three showed signs of failure by through-wall fracture. Four results have been obtained so far from the Health and Safety Laboratories (HSL), Buxton, where the tests took place.  A comparison of these results with those predicted numerically using the Johnson-Cook strength model used to simulate the pipe response is shown in fig. 7.

The hoop strains predicted by finite element simulation are a close approximation to the experimental values for the range of charge sizes used. Differences in the results may possibly be attributed to the parameters for the Johnson-Cook material model in AUTODYN 2D being those for 4340 steel [6] and differences may occur when the material is properly characterised by high rate testing. However, from the results obtained so far, it appears that the numerical solution gives a good indication of the final plastic hoop strains in the pipe walls.

An example of the deformed pipe can be seen in fig. 8 after explosive loading with 0.6 kg PE4.

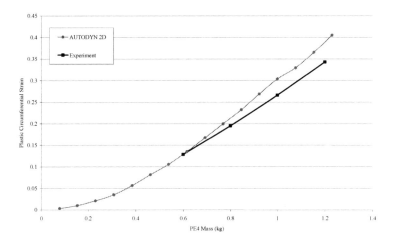

Figure 7:    A comparison between the numerical predictions using the Johnson-Cook material model and experimental results for the blast loaded pipes.

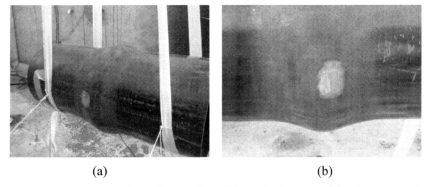

(a)                                              (b)

Figure 8:    Plastic deformation in the mild steel pipe caused by the detonation of the PE4 explosive.

In this test, large plastic deformation occurred without any wall failure. The bulging in the pipe is highly localised to the region where the cylindrical charge was placed and cracking in the pipe lacquer (fig. 8b) in close proximity to the bulging would indicate substantial longitudinal deformation.

## 6   Conclusions

In an attempt to explore the failure process of containment vessels subjected to internal explosive loading, a series of full-scale field tests on steel pipes combined with numerical and analytical modelling was conducted with funding from AWE. The results displayed in this paper are for those pipes that underwent plastic deformation with no wall failure. From the numerical analyses of the problem it was found that material behaviour and rate effects are dominant factors that affect the response of the vessel considerably. Hence, it is important to be able to define and model the material properties as accurately as possible in the simulation.

A comparison between the existing and derived theoretical analyses and the numerical modelling showed that the equation proposed by Duffey and Mitchell [1] was a closer approximation to the finite element simulation when determining the final plastic hoop strain in a blast loaded cylindrical shell.

## Acknowledgements

The sponsorship by AWE plc and EPSRC (DTA) has enabled this work to be carried out and is gratefully acknowledged.

## References

[1] Duffey, T. & Mitchell, D., Containment of explosions in cylindrical shells. *International Journal of Mechanical Sciences*, **15**, pp. 237–249, 1973.

[2] Fanous, F. & Lowell, G., Simplified analysis for impulsively loaded shells. *Journal of Structural Engineering*, **114(4)**, pp. 885–899, 1988.

[3] Clayton, A., Preliminary design of vessels to contain explosions, *Proceedings of 11$^{th}$ International Conference on Pressure Vessel Technology*, PVP2006-ICPVT11-93735, Vancouver, 2006.

[4] Baker, W., The elastic-plastic response of thin spherical shells to internal blast loading. *Journal of Applied Mechanics*, **27(1)**, pp. 139–144, 1960.

[5] ANSYS AUTODYN v.11 (2006) *Explicit Software for Non-linear Dynamics*. ANSYS, Inc., www.ansys.com.

[6] Johnson, G. & Cook, W., A constitutive model and data for metals subjected to large strains, high strain rates and high temperatures. *Proceedings of the 7$^{th}$ International Symposium on Ballistics*, The Hague, pp. 541–547, 1985.

[7] Departments of the US Army, Navy and Air Force. Structures to Resist the Effects of Accidental Explosions, *Report TM5-1300*, November 1990.

# Author Index

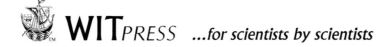

## Safety and Security Engineering IV

*Edited by:* **M. GUARASCIO**, *University of Rome 'La Sapienza', Italy;* **G. RENIERS**, *University of Antwerp, Belgium;* **C.A. BREBBIA**, *Wessex Institute of Technology, UK and* **F. GARZIA**, *University of Rome 'La Sapienza', Italy*

Safety and Security Engineering, due to its special nature, represents an interdisciplinary area of research and applications that brings together, in a systemic view, many disciplines of engineering, from the most traditional to the most advanced and novel.

Safety and Security Engineering is characterised by a totally new approach since it first analyses the hazard context, not only by means of traditional tools but also by means of risk analysis techniques, and then manages the above-mentioned context through technical solutions, installations, systems, human resources and procedures to prevent and face incidental events, natural and voluntary, that could damage people or goods.

The Fourth International Conference on Safety and Security Engineering (SAFE 2011) was convened to present and discuss the most recent developments in the theoretical and practical aspects of Safety and Security Engineering. The Conference papers in this volume cover the following topics: Infrastructure Protection; Risk Analysis, Assessment and Management; Public Safety and Security; Modelling and Experiments; Construction Safety and Security; Transportation and Road Safety; Safety of Users in Road Evacuation; Emergency and Disaster Management; Process Safety and Security; Emerging Issues in Safety.

*WIT Transactions on The Built Environment, Vol 117*
ISBN: 978-1-84564-522-9      eISBN: 978-1-84564-523-6
Published 2011  /  544pp  /  £234.00

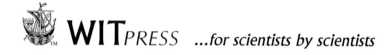

# System Identification for Structural Health Monitoring

*I. TAKEWAKI*, Kyoto University, Japan; **M. NAKAMURA**, *Technical Research Institute, Obayashi Corporation, Japan and* **S. YOSHITOMI**, *Kyoto University, Japan*

*System Identification for Structural Health Monitoring* is the first textbook on smart techniques of mechanical system identification using records from limited locations. The techniques explained in the book are based on rich content published in international journal papers by the authors, to which have been added introductory explanations to make the material accessible for a broad class of readers.

System identification (SI) techniques play an important role in investigating and reducing gaps between the constructed structural systems and their structural design models and in structural health monitoring for damage detection. A great amount of research has been conducted in SI.

There are two major branches of SI: modal-parameter and physical-parameter. The former is appropriate for identifying the overall mechanical properties of a structural system and exhibits stable characteristics in implementation. While the latter is important from different viewpoints, e.g. enhancement of reliability in active controlled structures or base-isolated structures, its development is limited due to the requirement for multiple measurements and the necessity of complicated manipulation. A mixed approach is often used in which physical parameters are identified from the modal parameters obtained by the modal-parameter SI. However, a sufficient number of modal parameters must be obtained in order for the unique and accurate identification of the physical parameters to take place. This requirement is usually hard to satisfy.

In spite of the importance of damping in the seismic-resistant design of buildings, it does not appear that its identification techniques have been developed sufficiently. Furthermore it is believed in general that the acceleration records for all the floors above a specific story are necessary in order to evaluate the story shear force that is required for stiffness-damping evaluation. This instrumentation may be unrealistic in multi-storied buildings.

To overcome this difficulty, the authors explain a unique system identification theory for a shear building model. They show that unique identification of story stiffnesses and viscous damping coefficients is possible when acceleration records at the floors just above and below a specific story are available.

ISBN: 978-1-84564-628-8     e-ISBN: 978-1-84564-629-5
Published 2011 / 272pp / £130.00

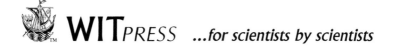 **WIT**PRESS   *...for scientists by scientists*

## Structures Under Shock and Impact XII

*Edited by: N. JONES, The University of Liverpool, UK and C.A. BREBBIA, Wessex Institute of Technology, UK*

Of interest to engineers from civil, military, nuclear, offshore, aeronautical, transportation and other backgrounds, this book contains the proceedings of a well-established biennial conference on the subject that was first held in 1989.

The shock and impact behaviour of structures presents challenges to researchers not only because it has obvious time-dependent aspects, but also because it is difficult to specify the external dynamic loading characteristics and to obtain the full dynamic properties of materials. It is crucial that we find ways to share the contributions and understanding that are developing from various theoretical, numerical and experimental studies, as well as investigations into material properties under dynamic loading conditions. This book helps to meet that need.

Topics covered include: Impact and Blast Loading Characteristics; Protection of Structures from Blast Loads; Energy Absorbing Issues; Structural Crashworthiness; Hazard Mitigation and Assessment; Behaviour of Steel Structures; Behaviour of Structural Concrete; Material Response to High Rate Loading; Seismic Engineering Applications; Interaction Between Computational and Experimental Results; Innovative Materials and Material Systems; Fluid Structure Interaction.

*WIT Transactions on The Built Environment, Vol 126*
**ISBN: 978-1-84564-612-7       e-ISBN: 978-1-84564-613-4**
**Forthcoming 2012  /  412pp  /  £172.00**

We are now able to supply you with details of
new WIT Press titles via
E-Mail. To subscribe to this free service, or for
information on any of our titles, please contact
the Marketing Department, WIT Press, Ashurst
Lodge, Ashurst, Southampton, SO40 7AA, UK
Tel:  +44 (0) 238 029 3223
Fax: +44 (0) 238 029 2853
E-mail:  marketing@witpress.com

Lightning Source UK Ltd.
Milton Keynes UK
UKOW032020190613

212535UK00001B/54/P